Atlas of Natural and Induced Fractures in Core

Atlas of Natural and Induced Fractures in Core

John C. Lorenz
FractureStudies LLC
New Mexico, United States

Scott P. Cooper
FractureStudies LLC
New Mexico, United States

WILEY Blackwell

AAPG
Advancing the World of Petroleum Geosciences™

This edition first published 2018

Registered Offices
John Wiley & Sons, Inc., 111 River Street, Hoboken, NJ 07030, USA
John Wiley & Sons Ltd, The Atrium, Southern Gate, Chichester, West Sussex, PO19 8SQ, UK

Editorial Office
9600 Garsington Road, Oxford, OX4 2DQ, UK

For details of our global editorial offices, customer services, and more information about Wiley products visit us at www.wiley.com.

Wiley also publishes its books in a variety of electronic formats and by print-on-demand. Some content that appears in standard print versions of this book may not be available in other formats.

Library of Congress Cataloging-in-Publication data applied for

ISBN: 9781119160007 [hardback]

Cover Design: Wiley
Cover Image: Photos by Scott P. Cooper

Set in 10/12pt Warnock by SPi Global, Pondicherry, India

C9781119160007_110723

Dedication

For my family, always there for each other, extraordinary.
Scott P. Cooper

For Bridget and Jackson, and the vast potential of a child's unknown future.
John C. Lorenz

Contents

Foreword

Fractured reservoirs are a significant part of our hydrocarbon production across the world, even though most were discovered to be fracture-dominated after initial drilling. These fractured reservoirs are some of the largest and most productive oil and gas fields worldwide. Over the last three decades, we have advanced in our understanding of these reservoirs and how to model the fracture systems quantitatively and create a Static Conceptual Fracture Model. Today, using the Static Conceptual Fracture Model as input in reservoir simulations allows us to explain previous production behavior and to predict future reservoir response. These models require compilation and integration of numerous data types derived from multiple observations and discipline sources.

This atlas by Lorenz and Cooper looks exclusively at one of these observational sources, that of core evaluation. For many studies, this is the "ground truth" of subsurface fracture characteristics. However, what we see in the core from the perspective of fractures is a mix of what is in the subsurface and what we have done to the core during and after the coring process. These "artifacts" must be omitted from our descriptions as not naturally occurring. I believe that a strong point of this atlas is in learning how to make this distinction.

Using the examples displayed in this volume, we can determine which fractures are naturally occurring and which are induced due to effects of coring, core handling, weathering, and stress relief, among others. The remaining fractures are considered natural fractures and their distribution and character can be interpreted to determine the causative forces that generated them. The reasons why we make these determinations are to predict fracture distributions and characteristics for future exploration and development wells and to constrain numerically a Static Conceptual Fracture Model in reservoir simulation.

Previous catalogs of fractures in rock have been published in book format by Kulander et al. (1990) and in numerous industry-related course notes covering fractured reservoirs by both Lorenz and Cooper and myself. However, this volume goes far beyond the previous versions in terms of completeness. It is a testament to the vast experience of the co-authors in working with fractures in core. I can think of none better to organize and create this atlas.

The authors have extensive experience over several decades in describing and quantifying fractures in outcrop and core. Much of this experience has been in direct support of oil and gas company operations and industry/academia consortia as well. Lorenz and Cooper have seen it all and understand how to interpret it.

I believe that this volume is the ultimate resource for "reading the story and history" of fractures in rocks from core. It is a "must-have" volume for all who have, or wish to have, an intimate knowledge of the rocks they work with from a fracture point of view.

R.A. Nelson
Broken N Consulting, Inc.

Preface

This book describes the characteristics of individual fractures as they are expressed in cores cut from hydrocarbon reservoirs, and it provides criteria for distinguishing the various fracture types as they are expressed in the limited fracture samples intersected by those cores. Different natural fracture types have different effects on reservoir permeability, and induced types have little or no effect at all, so it is imperative to correctly interpret cored fractures if the goal is ultimately to understand the broader fracture system and its influence in enhancing or degrading reservoir permeability.

The butts and slabs from a 30-ft long, 4-inch diameter core comprise a significant amount of material when laid out in the lab for detailed examination, but volumetrically this is a miniscule sampling of a reservoir. Thus it is imperative to maximize the amount of data obtained from the few fractures captured from such a small sampling of a reservoir, as well as to correctly interpret those fractures.

The fractures described here are the building blocks for the fracture networks and systems that influence reservoir permeability. Discussions and descriptions of those three-dimensional systems or of their behavior within a changing stress field as a reservoir is stimulated or produced are not part of this volume. Neither do we discuss fracture mechanics and some of the issues surrounding the origin of fractures, although some of our prejudices are evident.

Rather, we present here examples of the wide range of fracture types and characteristics that influence reservoir permeability so that they can be recognized in cores and used to construct both conceptual and numerical models of reservoirs.

Acknowledgments

We have had the privilege and challenge of assessing fractures in over 100 000 feet of core for numerous companies across the globe, and almost all of the companies have generously granted us permission to use photos of fractures in those cores as illustrations in this atlas. Permissions have been granted on the condition that the cores are not identified by well name, location, formation, or company name. This may remove some of the context for the fracture illustrations but in fact, most fracture types are not restricted to one formation or to one basin. Therefore, we have been happy to exchange constraints on core identification for the opportunity to select from a wide data set for this volume.

We have also spent time assessing cores at several state and federal core repositories where the data are in the public domain, we have worked for and in conjunction with various state geological surveys and university research groups, and we have included examples from those studies in this collection.

We whole-heartedly thank and lift our glasses in a toast to the following companies and government entities which have generously allowed us to use photos of their cores: Aera Energy LLC, Anschutz Exploration Corporation, ConocoPhillips Corporation, the Core Research Center of the US Geological Survey, Devon Energy Corporation, the Enhanced Oil Research Institute at the University of Wyoming, EOG Resources, Inc., Entreprise Tunisienne d'Activités Pétrolières, Gulf Keystone Petroleum International Limited, Hess Corporation, Kalegran B.V. (Member of the MOL Group), Kansas Geological Survey, Kinder Morgan Inc., Laredo Petroleum, Inc., Marathon Oil Corporation, Ministry of Natural Resources of the Kurdistan Regional Government Iraq, New Mexico Subsurface Data and Core Libraries part of the New Mexico Bureau of Geology and Mineral Resources at New Mexico Institute of Mining and Technology, Occidental Petroleum Corporation, Oklahoma Petroleum Information Center part of the Oklahoma Geological Survey at the University of Oklahoma, Petra Energia SA, Pioneer Natural Resources Company, Rocky Mountain Oilfield Testing Center, Samson Resources, Statoil ASA, Yates Petroleum Corporation, US Department of Energy, Utah Core Research Center part of the Utah Geological Survey, and the Wilson M. Laird Core and Sample Library part of the North Dakota Geological Survey at the University of North Dakota.

Introduction

Purpose of the Atlas

We were once emailed a long list of questions, arranged with paragraph-sized spaces below each question for our detailed answers and including photos of specific cored fractures, from a student starting to work on fractured cores. The questions were both basic and important, and included queries such as: Are both the slabs and butt of the cores used in fracture studies? How do you distinguish extension from shear fractures? Should I record the induced fractures? The list illustrated some of the problems and uncertainties in understanding natural fractures in core; it also indicated that people charged with assessing fractures in core do not always know enough about fractures, or cores, to make valid assessments.

This atlas is a tool, intended to help geologists recognize, differentiate, and interpret different types of natural fractures, induced fractures, and artifacts found in cores. We hope that this atlas will provide a reference for cored fractures for the industry, one that enables geologists to recognize the differences in fracture types as well as the significantly different effects that the different types have on a reservoir. Moreover, we hope that it fills what we perceive to be a gap in the literature, in that many fractured-reservoir textbooks start fracture analyses with the assumption that a geologist can already recognize and differentiate the various fracture types in a data set. We sincerely hope that this volume complements the seminal works of Nelson (1985, 2001) and Kulander et al. (1990).

The default concept of fractures is that they are planar, open cracks in a formation, when in fact there are many types of fractures and the different types can have significant differences in planarity, roughness, aperture, length, spacing, interconnectedness, and height, all affecting permeability. Knowledge of whether a fracture system formed in extension or shear, whether the fractures are open or mineralized, whether they are dissolution-enhanced slots or slickensided shear planes, all play into understanding the effects of fractures on a reservoir.

Significant information on a fracture network and the associated in situ stress system can often be found in a core even though core is a relatively miniscule and one-dimensional sampling of a reservoir. Since core is expensive and the sample is small, it is incumbent on the geologist to maximize the amount of fracture information recovered from a core, through knowledgeable examination of the core and informed analysis of the data collected from it.

We hope that this book provides a means for identification of many of the different fracture types found in cores as well as criteria for distinguishing between them. Some fracture types are widespread and control the basic plumbing of a reservoir, others are local and have minimal effect on formation permeability. A few fracture types provide orientation references or useful information on the in situ stress system. We have also illustrated some of the non-fracture artifacts found in cores since many of them provide context as well as important information that can be used for natural fracture analyses.

Scale of Interest

This book illustrates fractures at the scale of four-inch diameter cores. The illustrations are clarified where necessary with close-up photos, but for the most part we have not illustrated fractures at either the millimeter scale that is important to fracture mechanics, or at the meter/outcrop scale that is important to the construction of fracture-controlled permeability networks. This atlas is restricted to illustrating individual fractures as viewed when logging core, and describes their potential as individual permeability pathways. These structures and their basic interpretations are the primary building blocks for a complete fracture assessment and analysis, so they must be properly identified and correctly interpreted if subsequent analyses, interpretations, and modeling efforts are to be valid. For example, shear fractures commonly form intersecting conjugate pairs whereas extension fractures commonly form as single, parallel

Atlas of Natural and Induced Fractures in Core, First Edition. John C. Lorenz and Scott P. Cooper.

sets, and the difference greatly influences drainage and well-to-well interference patterns in a reservoir.

Industry geologists have recently had fewer opportunities to participate in the on-site coring process due to changing techniques, liability issues, and the increasing use of service companies to retrieve and process cores. Fewer company geologists have the opportunity to be familiar with drilling operations or with coring and core processing procedures; thus they are often unfamiliar with the important ways in which such operations affect a core. Geologists rarely get to look at a core any more until it has been cut, cleaned, marked, plugged, slabbed, boxed, sampled, and laid out in the lab, by which time significant natural fracture information has been lost and additional fractures have been created in the core.

Once the geologist gains access to a core, a big gap looms between counting and understanding fractures. It is easy to count fractures and measure their dips and strikes, which provides a data base that can be readily analyzed statistically. But such analyses are meaningless if fractures are not fully understood and fully characterized before they are analyzed, since fractures are so much more than planar breaks in the rock.

Core samples typically consist of fresh exposures of the rock and therefore provide unweathered detail compared to outcrops. However, the ability to extrapolate beyond the core into the other two dimensions in a reservoir is limited; for example, it is difficult to derive the lateral spacing of vertical fractures from the data provided by a vertical core unless the restrictive assumptions of Narr's (1996) analysis are met. Likewise, fracture heights are difficult to assess in horizontal core. Nevertheless, with experience and carefully acquired data, one can construct conceptual and often even semi-quantitative models of the three-dimensional fracture distributions, dimensions, spacings, and interconnectivities from cores.

Fracture Classification

Several systems have been used in classifying natural fractures. Some systems are based on fracture geometry, some on fracture origin, some on their electrical properties, and some on the potential effects of a fracture on a reservoir. For example, Nelson (2001) offers several classification schemes based on origin (extension, tension, or shear), on a fracture's potential permeability (open fractures vs. filled fractures), or the structural associations of the fracture system (fault related, fold related, regional, etc.). In contrast, petrophysicists commonly classify fractures in image logs by their electric or acoustic properties (i.e., "conductive" or "resistive").

For this atlas, natural fractures are divided into two main categories based on origin, i.e., extension fractures vs. shear fractures, with subcategories and modifiers for postfracture alterations.

It is human nature to categorize and classify, but as often as not, we are artificially compartmentalizing samples that form parts of a spectrum rather than discovering and documenting natural divisions. Fracture categorization serves a purpose but in fact, fractures may grade from one category into another. For example, "hybrid shears" (Hancock, 1986; Hancock and Bevan, 1987) offer a bridge between extension and conjugate-shear fracture categories, and induced petal fractures morph into and blend with centerline fractures. Natural fractures can also be reactivated over geologic time intervals, leading to ambiguities in classification, i.e., fractures that formed in extension are sometimes reactivated in shear. Similarly, the distinction between faults and shear fractures would seem to be self-explanatory, but if the distinction is based on offset magnitude it is arbitrary since shear offsets occur within a continuous range.

Organization of the Atlas

This atlas is organized into three parts.

Part 1: Natural Fractures

This section describes the characteristics of extension and shear fractures in core, which is not without complications. Most extension fractures are vertical, but intermediate-angle and horizontal fractures are also found in some cores. Shear fractures for the most part can be subdivided into Anderson's (1951) three dip-angle categories, corresponding to high-angle strike-slip shears, intermediate-angle dip-slip shears, and low-angle reverse dip-slip shears, but shear fractures with oblique slip and bedding-parallel slip are common in cores cut from some structural settings. We have also included short descriptions of other, less common types of cored natural fractures such as ptygmatically folded fractures and deformation bands.

Part 2: Induced Fractures

The two most important units in this second section on fractures created by coring and handling processes describe petal fractures, which can take many different forms, and centerline fractures. These two induced fracture types are important because they can be used to orient both a core and the natural fractures it contains relative to the in situ stress field, and sometimes even relative to north. Other induced fracture sections include descriptions of fractures created by twisting the core, by bending the core, and by dragging orientation scribe-knives along the core surface. Percussion-related fractures such as those created by enthusiastic rough-necks with hammers are also described.

Part 3: Artifacts

The third section of this book describes non-fracture structures created by coring and core-processing operations. The collection of these structures includes polished fracture faces in horizontal cores, core-catcher scars, and spinoffs. Some of these features are insignificant yet it is important to be able recognize them as such and to distinguish them from more important structures. Other artifacts are useful in assessing aspects of natural fracture systems in a core, or they may modify and obscure more important fractures, and many offer useful clues for reconstructing coring and processing operations.

We considered including "Miscellaneous" and "Problematica" sections for those fractures and features which are uncommon and/or difficult to interpret, but ran up against space limitations. We also contemplated a Part 4: Fracture Analysis Techniques, but that subject turns out to be broad enough to comprise a separate book.

Provenance of the Photographs

The photographs used in this atlas have been collected over the course of several decades while doing fracture analyses for different clients. Almost all of the client companies have generously given permission to use the photos (see Acknowledgments) provided that we do not identify the cores by formation, location, or company name. We have, however, provided the basic lithology of the cores, and in some cases their structural context.

Identification of the cored formations would have been of interest, but in fact the illustrations of fracture types and characteristics are largely independent of that information, and we would not have been able to use the photos if we had to identify them. We have not obscured the depth markings found in many of the core photos, but neither have we tried to indicate whether the indicated depths are in meters or feet.

Limitations of Using Photographs to Illustrate Fractures

Even though fractures are planar, it is necessary to assess them in three dimensions. Unfortunately, photos are two-dimensional views of fractures. The expression of a fracture on a core slab face or in a photograph commonly provides only an apparent dip, strike, width, and height. Multiple photos taken from different angles and with different lighting are usually required in order to convey the characteristics and subtleties of fracture parameters. Most of the photos included in this atlas are not formal, neatly framed photos of the core. Rather, they are candid photos taken from the angles that best illustrate the features of interest. Photos at different scales and taken from different angles are necessary for showing the characteristics of a fracture's surface, terminations, disposition relative to the host lithology, apertures, etc. Moreover, the relationship of one fracture to another requires photos from different angles to portray a true sense of the intersection geometry.

Most core libraries, warehouses, and laboratories have almost too much light for a good fracture analysis. Bright lights from multiple angles obscure the important but low-amplitude topographic textures that ornament many fracture faces. Fracture faces must be obliquely lighted in order to highlight these features, meaning that some of the photos that best illustrate fracture surfaces in this atlas are unevenly illuminated.

We initially tried to include examples of less than ideal fractures in this atlas, but it turned out that they are a waste of much-needed space. There is little advantage in showing the reader photos of poorly defined features and starting the caption with "If you could see it..." Thus we have primarily used the best photos of the best examples to illustrate the characteristics of the different fracture types, leaving it to the geologist working in the core warehouse to recognize the less than perfect examples using knowledge of what the more complete, more fully formed, or better exposed structures look like. We have, however, tried to illustrate the commonalities and important characteristics that define the range of characteristics of a given fracture group.

There was a synergy in organizing 9,000–10,000 photos and culling them to meet the limitations of this publication in that for the first time we were seeing all of the examples together, affording an opportunity to compare and contrast. The exercise of sorting the myriad photos by fracture type imposed a certain organization on the data set and brought some concepts into better perspective. One concept was the recognition that the ranges of characteristics defining some fracture types are wide enough to overlap with the ranges of other fracture types. Moreover, it forced us to decide what features of each fracture type are important and diagnostic.

We have not provided an exact scale for each photo, but we have indicated the core diameter in each caption. Additionally, most of the photos show a pencil point or a finger, intentionally included in the photo, as a convenient if approximate scale, and more easily set up than a more quantitative scale when taking hundreds of photos in the course of a single core analysis.

Core Marking Conventions and Terminology

As much as possible, we have oriented photos to present cores in the in situ position, i.e., the top of the photo is oriented in the stratigraphic up direction, and the long axes of vertical cores are vertical whereas the long axes of horizontal cores are horizontal. The core orientations are noted in each photo caption. There are exceptions for photos of long core intervals, and for photos of artifacts created when the long axis of a core was horizontal when the artifact formed, for example during slabbing, regardless of its in situ position.

There are several conventions for marking a core's in situ position, the most common one being a pair of parallel lines, one red and the other black, drawn along the core axis on the outer core surface, with "red on the right looking uphole." Uphole is also usually the stratigraphic up direction (Figure 0.1) for vertical cores. There are color variations on this convention, although most companies at least use a red line on the right.

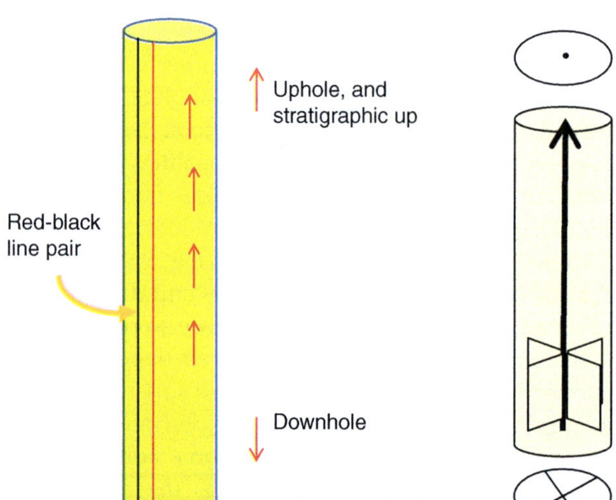

Figure 0.1 Left: the common red–black line pair drawn on a core surface indicates the uphole direction, with "red on the right looking uphole". This color combination is typical but not universal. Some cores are marked with up arrows instead of red–black line pairs; more rarely, the arrows are drawn to point in the downhole direction. Right: the orientations of some core cross-sections are annotated for photography with circled dots and circled "Xs", representing the up arrows as viewed from the front and from the rear, respectively.

Arrows may also be used to indicate core orientation, with the arrows typically but not universally pointing uphole. A less common but usually recognized convention is to indicate that a view of the end of a core is downward by drawing a circle around a dot on the cross-section core surface, the dot representing the point of the uphole arrow viewed from the front, i.e., with the arrow pointing at the viewer while looking downhole. Likewise, an uphole view of a core cross-section may be represented by a circle drawn around an X, representing a view of the back or feathered end of the upward-pointing arrow.

Readers will note that mixed English and metric units are used throughout the atlas. Many of the core depths and diameters were originally designated in feet and inches, respectively, and they have not been changed since these units are still widely used and understood in the oil industry. However, other dimensions and measurements that are not related to coring and drilling processes, such as fracture widths, are given in the metric system, using millimeters and centimeters.

The terminology used for cores cut from vertical holes is generally unambiguous, with the uphole direction aligned with vertical and also with stratigraphic up unless the core cut tilted bedding. In contrast, cores cut from deviated and horizontal wells provide several opportunities for misunderstandings since the uphole direction is no longer equivalent to vertical. Red–black line pairs are also used on horizontal cores, but the "uphole" direction is not stratigraphic up. "Uphole" for a horizontal core refers to the direction towards the heel of the well, the bend where the well turns from horizontal to vertical around the build angle (Figure 0.2). The downhole direction is towards the toe of the well at the limit of drilling. Stratigraphic up, where it can be determined in a horizontal core, is typically referred to as the high side or dorsal side of the core, in contrast to the low side or stratigraphically lower side of the core.

When oriented, a horizontal core is oriented relative to vertical rather than relative to north, and the position of the principle scribe line is given in the core orientation report in degrees of rotation from the high side of the core when looking downhole. If a horizontal core is not oriented, the dip of horizontal bedding in the core can often be used to determine the high side. There are two possible high-side determinations for unoriented core that cuts parallel to bedding, and the ambiguity cannot be resolved unless sedimentary geopetal structures are present. If a core is not oriented and does not display bedding, the high side of the core usually cannot be determined.

As yet there is no industry-wide convention, but the high or dorsal sides of horizontal cores are often marked with a blue line. However, the high side of a core can easily be misidentified, and where present, such high-side markings have not always been made with a full

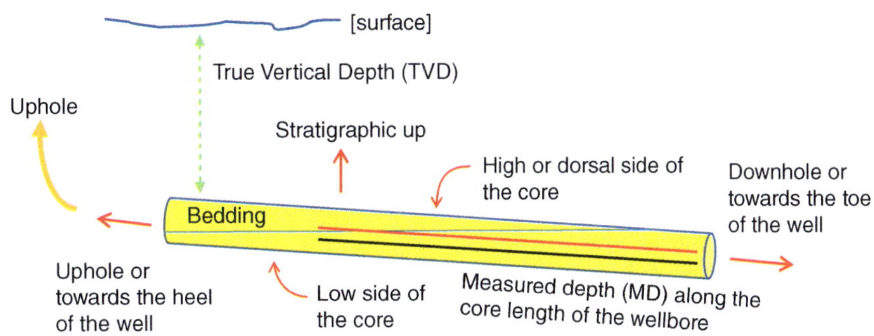

Figure 0.2 Where it can be determined, the stratigraphic top of a horizontal or near-horizontal core is referred to as the high side or dorsal side of a core and is different from the uphole direction, toward the heel of the well, that is indicated by the red–black line pair drawn on the core surface. Another line, commonly blue, may be drawn on the high side of a core where this determination can be made, but there is commonly ambiguity in making the determination and the line should be used with caution.

understanding of local bedding dip and wellbore deviation. High-side markings on a horizontal core should be used with caution.

Definitions

A division exists between geologists who prefer the term "joint" and those who prefer the term "fracture" to refer in general to breaks in a rock. Those who prefer "joint" use the term "vein" to refer to mineralized joints, whereas those who prefer "fracture" refer to the same features as "mineralized fractures" and reserve "vein" for other applications. The division often falls along academic versus industry lines, although even joint-oriented geologists sometimes use terminology centered on "fracture" in conversation and informal settings.

"Fracture" is the standard if informal term used by industry, and we prefer it for most applications since it is easy to append modifiers such as "induced," "extension," and "mineralized" to the term. It is also easy to string modifiers together (i.e., "high-angle shear fracture"), to create a flexible, widely applicable, and readily understandable lexicon. "Fracture"-based terminology is therefore used in this atlas although we have also found the terms "fissure" and "vein" useful.

The main drawback to the term "fracture" is that the engineering community also refers to the hydraulic stimulation injections used to enhance production as "fractures," sometimes leading to the convenient but inaccurate conclusion that stimulation fractures and natural fractures form in the same geomechanical manner.

In this atlas, we use the following definitions.

Fractures

Fracture: Any break in the rock, whether natural or induced, mineralized or unmineralized.

Natural fracture: Fractures created by geologic forces and processes.

Induced fracture: Fractures created by forces associated with the drilling, coring, handling, and processing procedures.

Extension fracture: A fracture, natural or induced, created within a system of three unequal compressive stresses, with the fracture plane oriented normal to the least compressive stress, and with the direction of opening normal to the fracture walls. These are also sometimes called "Mode I" fractures. They are not to be confused with true tensile fractures, formed where the rock is in tension in at least one axis and which are not common in subsurface geology.

Shear fracture: A fracture created by three unequal compressive stresses and oriented at an oblique angle, ideally 30°, to the maximum compressive stress. The opposing fracture faces are offset relative to each other in a direction parallel to the fracture plane. These are commonly referred to as "Mode II" fractures. Most shear fractures in cores are natural.

Crack: An unmineralized, narrow break in the rock along which the rock has not completely separated. The opposing faces are still attached to each other by local millimeter- to centimeter-scale webs of intact rock spanning the fracture. Cracks are commonly precursors to full separation of the rock across the fracture face.

Fracture Parameters

Fracture width: The linear distance between and normal to the host rock walls. Widths are useful in calculating percent strain in extension, but are only secondarily useful in assessing fracture permeability since the width may be occluded by mineralization. A fracture may be completely occluded by mineralization but still have width. The widths of extension fractures are commonly relatively uniform, narrowing only near the fracture terminations, whereas the widths of shear fractures are typically irregular. We record extension fracture widths as the maximum sustained width, and shear fracture widths as an estimated average width.

Fracture aperture: The linear dimension of the open void space between opposing fracture walls. A fracture aperture equals its width if it is unmineralized. A completely mineralized fracture has width but no aperture. Aperture controls fracture permeability, but apertures are not always uniform. Extension fractures that have undergone dissolution and/or mineralization, and most shear fractures, have irregular apertures. The difficulty comes in characterizing irregular apertures with one length measurement. Apertures in fractures with crystalline mineralization commonly resemble the irregular open void spaces between clenched teeth, through which fluid will flow but which cannot realistically be characterized by a single width measurement. Thus we commonly record remnant fracture porosity (see the next definition) for mineralized fractures.

Remnant fracture porosity: The percentage of the original fracture width that is still open and porous despite mineralization, i.e., that part of the fracture that would still be effective in storing and conducting fluid. This can be semi-quantitatively estimated by using percentage charts such as those published by Compton (1985) for estimating heavy mineral percentages. An unmineralized fracture retains 100% of its original width as void space regardless of its width. A fracture that has been completely occluded by mineralization has 0% remnant porosity regardless of its width. Most mineralized fractures retain some percentage of the original width as remnant open void space. Even fractures that appear to be completely occluded under a hand lens may still be more permeable than the microdarcy- or nanodarcy-scale permeability of the host rocks (Lorenz et al., 1989, 2005).

The engineering tradition has been that fracture permeability is proportional to the cube of a fracture's width (Warren and Root, 1963), but as pointed out by Wennberg et al. (2016), flow along a fracture is more commonly channeled flow around fracture irregularities than it is sheet flow between parallel plates.

Fracture bulk porosity: The contribution of fracture void space to the total/bulk volume of the rock, measured in percent. Even in highly fractured reservoirs, this is typically less than 1% and rarely as much as 2% (Nelson, 2002).

Other Fracture Characteristics

Fractography: The systematic ornamentation found on the faces of many fractures, useful in determining whether the fracture originated in extension or shear. Typical fractographic markings include plume structures and arrest lines on extension fractures, and steps or lineations on shear fractures. Fractographic markings do not form on all fractures and they may be obscured by mineralization, thus many fracture faces are nondescript and can be characterized only with descriptions such as "rough," "planar" or "undulatory." Other fracture surfaces show evidence for dissolution that has destroyed whatever fractographic markings may have existed.

Systematic: A term referring to a degree of organization and similarity in the characteristics of a system of fractures, i.e., fractures with similar strikes, dips, and surface features that are relatively evenly distributed in the rock and that can be differentiated from the fractures of other sets. Not random.

Fracture strike relative to other natural fractures: The measured intersection angle between two fractures within one piece of core, or the angle between projections of the fracture planes to an inferred intersection outside the core volume. This measurement provides an estimate of the degree of interconnectivity of a fracture permeability network. A 0° intersection angle indicates parallel fractures.

Fracture strike relative to induced fractures: The intersection angle between a natural fracture and a nearby petal or centerline fracture. Since these two types of induced fractures record the orientation of the in situ maximum horizontal compressive stress, this measurement provides an indication of the probable behavior (closure, shear, or minimal response) of the fractures during stress changes caused by reservoir production. Where present, these two induced fracture types provide a consistent orientation reference in a core.

Fracture-Related Structures

Fissures: Roughly planar, slot-like features with highly irregular widths suggestive of large-scale dissolution. They are commonly filled with poorly sorted exotic materials but may also be filled with locally derived materials of more uniform composition.

Veins: A variety of planar fractures of uncertain origin. They are typically completely filled with an amorphous, non-crystalline mineralization composed of material chemically similar to that of the host rock, suggesting formation and mineralization early in the sedimentation, burial, and diagenesis history of the rock. Many veins have embayed walls consistent with dissolution at some point in their history. Other veins are elliptical in cross-section, suggesting that the host strata were poorly lithified at the time of vein formation.

References

Anderson, E.M., 1951, The Dynamics of Faulting and Dyke Formation with Applications to Britain. Edinburgh, Oliver and Boyd.

Compton, R.R., 1985, Geology in the Field. Chichester: Wiley.

Hancock, P.L., 1986, Joint spectra, *in* Nichol, I., and Nesbitt, R.W., eds, Geology in the Real World – the Kingley Dunham volume. London: Institution of Mining and Metallurgy, p. 155.

Hancock, P.L., and Bevan, T.G., 1987, Brittle modes of foreland deformation, *in* Coward, M.P., Dewey, J.F., and Hancock, P.L., eds, Continental Extension Tectonics. Geological Society Special Publication, v. 28, pp. 127–137.

Kulander, B.R., Dean, S.L., and Ward, B.J., 1990, Interpretation and logging of natural and induced fractures in core, AAPG, Methods in Exploration Series 8.

Lorenz, J.C., Warpinski, N.R., Branagan, P.T., and Sattler, A.R., 1989, Fracture characteristics and reservoir behavior in stress-sensitive fracture systems in flat-lying formations. Journal of Petroleum Technology, 41, 614–622.

Lorenz, J.C., Krystinik, L.F., and Mroz, T.H., 2005, Shear reactivation of fractures in deep Frontier sandstones: Evidence from horizontal wells in the Table Rock Field, Wyoming, *in* Bishop, M.G., et al., eds, Gas in Low Permeability Reservoirs of the Rocky Mountain Region: Rocky Mountain Association of Geologists guidebook, pp. 267–288.

Narr, W., 1996, Estimating average fracture spacing in subsurface rock. AAPG Bulletin, 80, 1565–1586.

Nelson, R.A., 1985, The Geologic Analysis of Naturally Fractured Reservoirs. Houston: Gulf Professional Publishing.

Nelson, R.A., 2001, The Geologic Analysis of Naturally Fractured Reservoirs, 2nd edn. Houston: Gulf Professional Publishing.

Warren, J.E., and Root, P.J., 1963, The behavior of naturally fractured reservoirs. Society of Petroleum Engineers Journal, 228, 245–255.

Wennberg, O.P., Casini, G., Jonoud, S., and Peacock, D.C.P., 2016, The characteristics of open fractures in carbonate reservoirs and their impact on fluid flow: a discussion. Petroleum Geoscience, 22, 91–104.

Part 1

Natural Fractures

Section A

Extension Fractures

A1

High-Angle Extension Fractures

A1a Introduction

High-angle fractures (Figures A1a.1, A1a.2) are vertical or near-vertical, and are the most common type of fracture in simple and even in some complex structural settings. Many geologists and most modelers consciously or unconsciously default to extension fractures as an assumed universal fracture type.

Figure A1a.1 High-angle extension fractures may be present in a vertical core for many feet along the core axis if the core cuts through a homogeneous lithology, and if there was enough strain energy during fracturing to create large fractures. This fracture is mineralized with calcite but the core has split along the fracture, indicating that the mineralization is weak. The calcite mineralization is white but it is obscured by drilling mud that has invaded the incompletely mineralized fracture. The core is cradled in half of the aluminum core barrel liner, which has been split lengthwise with a saw in order to remove the core from the liner. Full-diameter/unslabbed 5¼ inch vertical core, fine-grained sandstone. Uphole is towards the top of the photograph.

Figure A1a.2 Many high-angle extension fractures are relatively small and obscure, especially as exposed on the rough outer surface of a core. This calcite-mineralized fracture, highlighted by dashed lines drawn parallel to the fracture trace with a silver marker, is largely intact, indicating relatively tight mineralization. The mineralized fracture face is broken open and exposed at the thumb. The fracture is planar, but its subtle trace on the core surface appears to be irregular due to the rough core surface. The fracture has a relatively uniform width but narrows and pinches out abruptly at the top and bottom where it terminates against less calcareous layers. This fracture would have been missed if the core surface had not been cleaned while logging. Full-diameter/unslabbed four-inch vertical core, calcareous shale. Uphole is towards the top of the photo.

Atlas of Natural and Induced Fractures in Core, First Edition. John C. Lorenz and Scott P. Cooper.

Extension fractures form as sets of parallel planes and are typically strata-bound. Therefore, assuming that fluid will flow more readily through the fractures than through the matrix (not always the case), a single set of extension fractures can significantly enhance the system permeability of a reservoir, but only in the horizontal plane and only in one direction. In order to fully assess the effects of fractures on reservoir permeability, however, the detailed fracture characteristics including fracture heights, terminations, spacings, strikes, distributions with respect to lithology, width, and mineralization as it affects fracture aperture must be assessed.

Extension fracture permeability can be dynamic since fracture apertures can close as fluid pressure within the apertures is preferentially reduced over matrix pressures during reservoir production. This effect is more prominent for narrow fractures where small amounts of closure constitute a significant percentage of the fracture aperture. Fracture characterization is especially important where two or more sets of oblique-striking extension fractures are present in a reservoir since one set may be less mineralized or more optimally oriented relative to the in situ stresses, and therefore more permeable than the other(s) even if it is less well developed.

A1b Fractography of High-Angle Extension Fractures

The process of fracturing commonly creates distinct low-relief patterns or "fractography" on a fracture face, and different fractographic patterns are diagnostic of an origin in shear vs. extension. The fractography of fractures has been described since the early days of geology, but it was only with the development of materials science and studies of the breakage patterns in ceramics and glass in the laboratory that the significance of these surface markings began to be understood.

Extension fracture surfaces are commonly decorated by plumes, arrest lines, and twist hackles (Figures A1b.1–A1b.3). As the name implies, plume or plumose structure is a subtle feather-like pattern that ideally has an axis like the shaft of a feather, with low-relief branches radiating away from the axis. Some plume structures, however, are less symmetric and wander less definitively across the fracture face. In core, many plumes are less than symmetric because the core has sampled only a small part of the fracture face and a small section of the plume pattern.

Plumes are usually inferred to indicate rapid fracture propagation. Plume axes record the direction of fracture propagation and can sometimes be traced back to the point of origin of the fracture, commonly a fossil, clast, or other heterogeneity that acted to concentrate stress and initiate fracturing. In some formations plumes originate at bedding contacts, in other formations the plumes originate at inhomogeneities within beds.

Arcuate arrest lines form ridges oriented normal to the branches of the feather structure, and record interruptions in the propagation of a fracture as well as the location of the fracture tip at the time the fracture growth was interrupted (e.g., Kulander et al., 1990).

Extension fracture edges may be marked by twist hackle, commonly interpreted to indicate a change in stress conditions near the contact of the host rock with another layer, where the change in mechanical properties altered the in situ stress conditions slightly. The propagating fracture plane changed strike, splitting into segments to do so. The en echelon pattern of twist hackle exposed in a 2D plane normal to the fracture can resemble the en echelon veins and steps common to some shear fractures, and care must be taken not to infer shear from an en echelon geometry alone.

The axis of a plume structure is usually considered to have been controlled by the maximum compressive stress at the time of fracturing. However, if the fracture

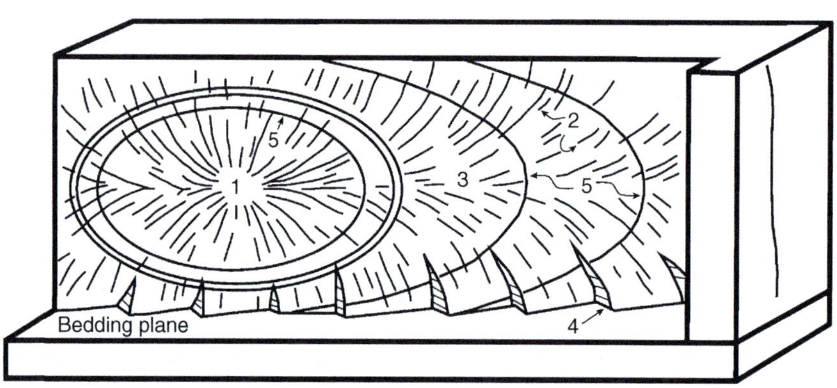

Figure A1b.1 Diagram of the ideal fractography of an extension fracture. Adapted from Kulander et al. (1990). 1 = Origin, 2 = Branches of the feather texture, 3 = Plume axis, 4 = Twist hackle, 5 = Arrest lines. Four-inch diameter core samples a relatively small segment of the fracture plane and must be examined carefully in order to recognize these elements.

Figure A1b.2 A strata-bound fracture decorated with a faint plume structure that propagated from left to right, and that broke into twist hackle in the altered stress areas near the upper and lower bedding contacts.

Figure A1b.3 Extension fracture in a sandstone showing an origin near the end of the hammer handle, possibly at the cross fracture, and an initial circular shape where the fracture radius was less than the thickness of the host bed. The fracture subsequently propagated incrementally, both left and right, pausing long enough in several places to create arcuate arrest lines. The plume pattern is normal to the arrest lines.

develops in horizontally layered strata, the axis of a strata-bound fracture commonly turns to follow horizontal bedding as the fracture grows, even if the maximum stress is vertical.

Plumes indicate extension, but not all extension fractures display plume structures. Plumes are most common in fine-grained and well-cemented rock, and they are rarer in coarse-grained rocks where the grain size exceeds the topographic relief of the plume. Plumes may not have formed where fractures propagated slowly, they are easily obscured by fracture-face mineralization, and they are often removed by fracture-face dissolution and by recrystallization during diagenesis, especially in limestones.

The small size of a fracture sample provided by a core can make it difficult to recognize extension fractures in the absence of plumes. Extension fractures can often be recognized by a lack of offset parallel to the fracture walls, easiest to determine where the fractured strata are finely bedded, but the possibility that offset was strike-slip, parallel to bedding, must also be considered where offset bedding is not obvious. Moreover, the magnitude of offset along shear fractures can be minimal, millimeter in scale, so care should be taken when examining core for offset.

A1b1 Plume Structure

Figure A1b1.1 Plumes with horizontal axes, recording left-to-right propagation, decorate the faces of several small strata-bound, closely spaced, unmineralized extension fractures. Rough surfaces at the top and bottom of the fracture indicate the contact with the untractured over- and underlying beds. Vertical four-inch diameter core from siltstone, uphole is towards the top of the photo.

Figure A1b1.2 Plume structure on the face of a vertical, bed-normal extension fracture in shale. The plume axis is horizontal, about a quarter of the way up from the bottom edge of the core, and indicates that the fracture propagated parallel to bedding, from left to right. The plume and fracture extend the full height of a calcareous bed in a cored shale. The upper and lower contacts of the bed, and the edge of the fracture, are indicated by increasing roughness of the fracture surface near the top and bottom of the photo, probably a minor form of twist hackle. Vertical four-inch diameter core, uphole is towards the top of the photo.

Figure A1b1.3 A large-scale plume records the propagation of this fracture face from right to left across the core, which was cut from a non-calcareous shale. The plume is without a well-defined axis but propagated upward and to the left in the upper half of the cored fracture, and downward and to the left in the lower half. The fracture is lightly mineralized with calcite but the plume relief is greater than the thickness of the layer of calcite and the ornamentation is apparent through the mineralization. Vertical four-inch diameter core, uphole is towards the top of the photo.

Figure A1b1.4 This core captured part of a plume that was much larger than the core. The fracture is mineralized with a light layer of calcite. Remnants of drilling mud (tan) stain the fracture face. Vertical four-inch diameter core, uphole is towards the top of the photo.

Figure A1b1.5 A subtle plume propagated horizontally from right to left across this fracture face. The strata-bound fracture, in a non-calcareous siltstone, is covered by a layer of diagenetic clay half a millimeter thick. The impression of the plume from the missing half of the fracture is preserved on the surface of the clay. Vertical four-inch diameter core, uphole is towards the top of the photo. The stump of an unsuccessfully recovered one-inch diameter porosity/permeability plug remains in the hole cut into the core.

Figure A1b1.6 Opposing faces of an extension fracture in limestone show mirror images of the plume structure on the two surfaces. The fracture grew horizontally in this rock. The narrow fracture is lightly mineralized with calcite on both faces. Vertical four-inch diameter core, uphole is towards the top of the photo.

Figure A1b1.7 A plume with a horizontal axis, highlighted by oblique lighting, records the propagation of an extension fracture from left to right in a fine-grained limestone interbedded in a shale sequence. The top edge of the fracture is marked by minor twist hackle where it approaches a shale layer. The bottom edge of the natural fracture terminates at another bedding plane but without twisting. The fracture plane was extended downward along the rougher curved surface during core processing, exiting the core towards the viewer. The vertical line in the fracture face is the slabbing saw cut. Vertical three-inch diameter core, uphole is towards the top of the photo.

Figure A1b1.8 A wandering plume structure on the face of an extension fracture propagated upward, oblique to the core axis in a limestone. Plume axes parallel to the core axis suggest coring-induced fractures, but this fracture is natural as suggested by the upward vs downward growth of the plume and its asymmetry relative to the core axis. The fracture terminates at the top at a shalier unit. Vertical four-inch diameter core, uphole is towards the top of the photo.

A1b2 Twist Hackle

Figure A1b2.1 Definitive twist hackle growing out of a poorly defined left-to-right plume structure, marking the changed mechanical properties and stress conditions near the bedding contact and the upper edge of the fracture. Vertical three-inch diameter core, uphole is towards the top of the photo.

Figure A1b2.2 Two short, parallel, laterally offset, strata-bound extension fractures in a strongly bedded siliceous shale. A subtle plume structure marks the lower fracture but none is evident on the upper fracture. The upper fracture, however, is marked by well-developed, low-relief twist hackle along both the upper and lower edges. Vertical three-inch diameter core, uphole is towards the top of the photo.

Figure A1b2.3 A fracture plane in a siltstone, without an obvious plume, twisting in the upper half of the core into hackles where the homogeneous rock changes to bedded rock. The twist hackles suggest different mechanical properties and slightly oblique stress regimes in the two zones. Vertical four-inch diameter core, uphole is towards the top of the photo.

Figure A1b2.4 Opposing faces of a vertical extension fracture in oil-stained sandstone that show twist hackle transitioning upward into a more planar extension fracture. The fracture face is not marked by plume structure since the sandstone is poorly cemented and not mechanically prone to the formation of plumes. Vertical three-inch diameter core, uphole is towards the top of the photo.

A1b3 Arrest Lines

Figure A1b3.1 Two views of an arrest line on a calcite-mineralized fracture face in shale. The left photo shows the opposing, calcite-mineralized faces of an extension fracture (opened in butterfly fashion along a horizontal axis) to show the arcuate rib of the arrest line. Note the complimentary patterns of mineralization on the two faces, adhering to one or the other face but not both. The right photo shows one face of this fracture plane edge-on, looking downhole to highlight the low-amplitude relief in the fracture face at the rib (*red arrow*), with the opposing face turned parallel to the plane of the photo. Vertical four-inch diameter core, uphole is towards the top of the left photo, and facing the viewer in the right photo.

Figure A1b3.2 Two views of a calcite-mineralized composite extension fracture in shale. The arcuate edge of the original fracture cuts approximately down the middle of the core, separating the planar surface of the original fracture (to the right in both photos) from the twist hackle surface marking a later extension of the fracture. Drilling mud obscures the fracture face. Twist hackles and the change in strike indicate a changed stress regime at the time of fracture extension. Vertical four-inch diameter core, uphole is towards the top of both photos.

A1c Extension Fracture Dimensions

Height, width, length, and spacing are some of the more important fracture parameters for estimating and modeling the influence of fracture permeability on a reservoir.

Fracture *widths* can be measured in intact cores and even if the core has broken open along a fracture, its width can at least be estimated. One train of thought suggests that fracture widths in core are inaccurate due to the expansion of a core after it has been cut from the rock and released from the in situ confining stresses. However, anelastic strain recovery techniques, developed to measure stress-release expansion in order to calculate in situ stresses (e.g., Teufel, 1983; Warpinski et al., 1993), suggest that lateral core expansion, and thus associated changes in fracture widths in intact core, are of the order of a few hundreds of microstrains (a few hundred parts per million) and are negligible. In fact, the thermal *contraction* of a core as it cools from reservoir to surface temperatures commonly has a greater effect on core diameter.

The *spacings* of high-angle fractures can sometimes be estimated from vertical cores, (e.g., Narr, 1996) but spacing is more reliably assessed in core from deviated wells where actual spacings can often be measured.

Measurements of the *heights* of high-angle extension fractures can be problematic in any core; full fracture height will be captured by vertical cores only if the fracture plane is parallel to the core axis, and will be almost impossible to assess from horizontal cores where the maximum measurable heights are limited by the core diameter.

Similarly, it is virtually impossible to obtain fracture *lengths* from the one-dimensional data acquired from either vertical or horizontal wells. Lateral fracture edges are occasionally captured by cores, but useful data on fracture lengths are best gathered from outcrops.

Extension fracture heights, lengths, and spacings are illustrated in this section of the atlas, along with vertical and lateral fracture terminations. Effective fracture widths are strongly influenced by mineralization and so they are addressed in the Mineralization section. Fracture strikes are also important and some are illustrated here, but strikes are addressed more completely later in the volume.

Where a robust set of core measurements for extension fracture dimensions can be obtained, the data for heights, widths, and spacings (and in outcrop, the fracture lengths) typically have lognormal distributions with a wide range of values. It is probably inaccurate to use single values for any of these dimensions as inputs for fracture models. The lognormal distributions include numerous short ("short" in terms of both height and length), narrow, and closely spaced fractures, with increasingly fewer taller, longer, wider, and more widely spaced fractures.

Similar parameters for shear fractures measured in cores are typically more irregularly distributed since cores capture such a small sampling of these less predictable fracture systems.

A1c1 Extension Fracture Heights and Vertical Terminations

Extension fracture height measurements are commonly limited by bedding, being controlled as much by the sedimentology of a formation, i.e., the differences in the mechanical properties of adjacent beds, as by the stresses that caused fracturing. Another consideration for height measurements is that for many taller fractures, the measured heights are commonly only minima since only part of a fracture height can be measured if it is inclined relative to the core axis and exits the side of the core before terminating (Figure A1c1.1). Taller fractures, whether extension or shear fractures, commonly exit the side of a core before terminating: if a core is four inches in diameter and contains a four foot tall fracture that is inclined 5° to the core axis, the fracture may exit the core top and bottom on opposite sides of the core. Measurable height is a minimum even if one of the two vertical fracture terminations is present in the core. Nevertheless, a core-based fracture-height data set (Figures A1c1.2, A1c1.3) is still useful as representative of the lower end of the range of fracture heights in a system, and even truncated data sets provide at least minimum values that can be used to estimate the true range of fracture heights.

It is also useful to log the thickness of the bed that hosts fractures and the location and nature of the fracture terminations in order to describe the effect of fractures on vertical system permeability. Cross-plots typically show a strong correlation between extension fracture heights and bed thickness where fracture height is controlled by mechanical stratigraphy. In contrast, extension fractures may cut indiscriminately across bedding where formations have a low degree of mechanical contrast across bedding. In these formations one should log both the locations of fracture terminations and the number and types of beds crossed by fractures.

Vertical fracture terminations, top and bottom, can usually be assigned to a few generalized categories including the following:

Known Terminations

- Termination at a lithologic, geomechanical boundary, suggesting a difference in the strain response of the two lithologies. Fracture terminations at boundaries may be abrupt, tapering, or splayed, but they suggest

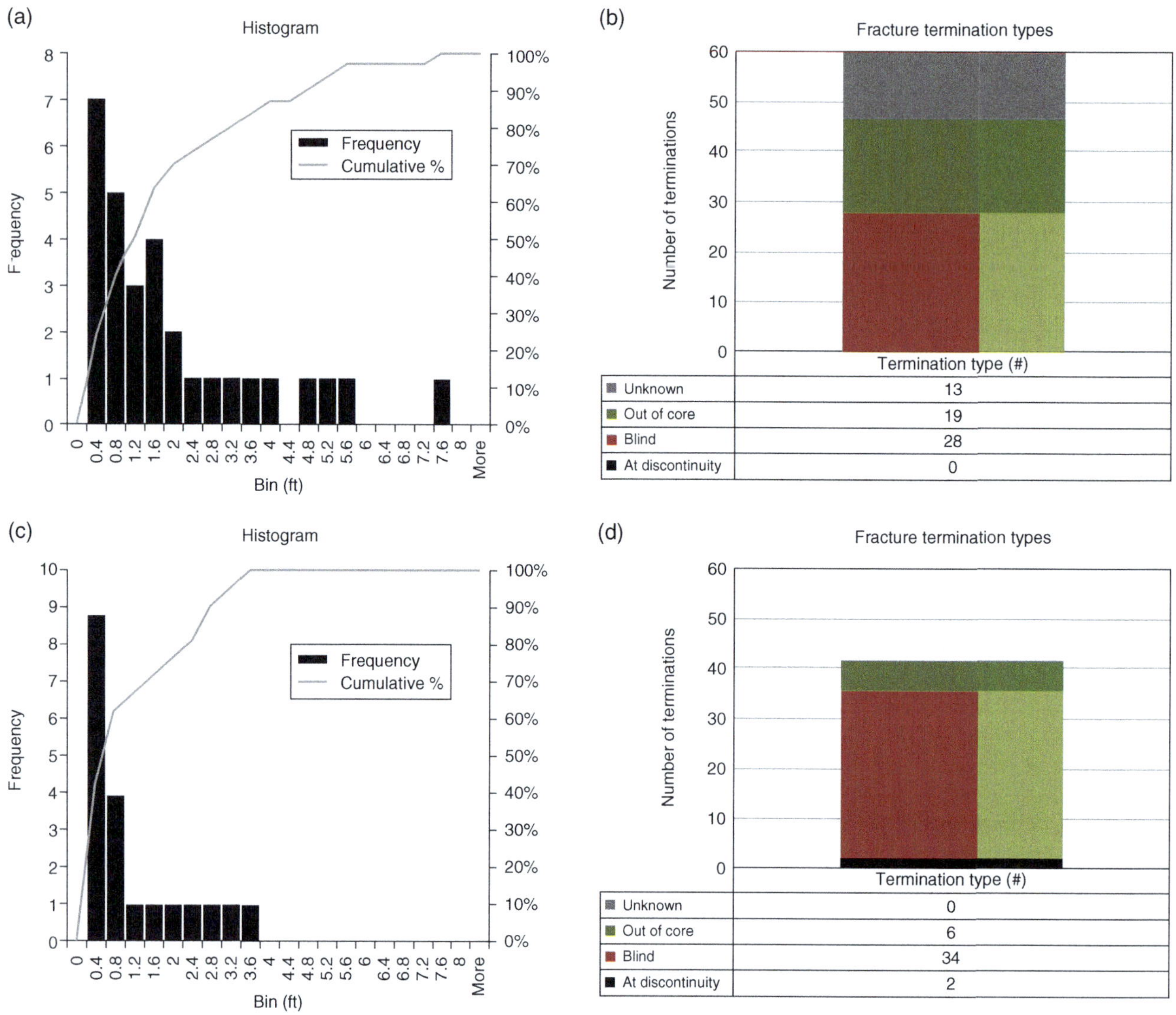

Figure A1c1.1 Histograms of the distributions of high-angle extension fracture heights and their terminations in cores from two vertical wells in the same poorly bedded marine shale formation. Top: fracture heights range from 0.1 to 7.60 ft, averaging 1.85 ft (n = 30 fractures with 60 terminations). Bottom: fracture heights range from 0.1 to 3.3 ft, averaging 1.12 ft (n = 21 fractures with 42 terminations). Most of these fractures terminate blindly within homogeneous rock, few terminate at bedding planes.

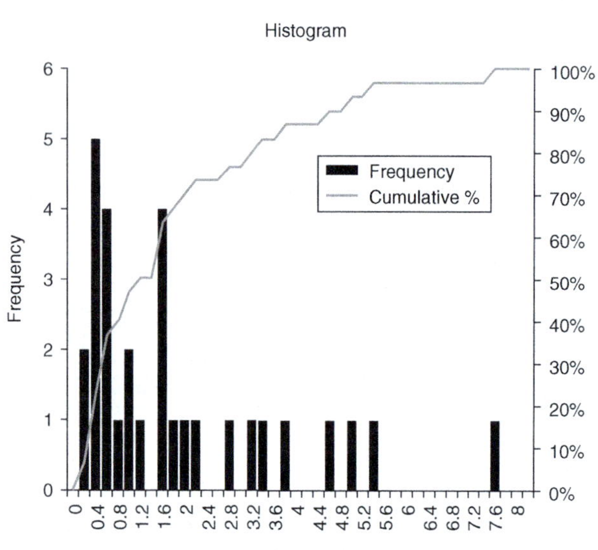

Figure A1c1.2 Height distribution of the same high-angle extension fracture data set used in Figure A1c1.1 upper left, but plotted here with 0.2 ft height bins rather than 0.4 ft bins. Height distributions that fall off abruptly to the left, with fewer smaller fractures, can be hidden if a relatively large bin size is chosen for the plot or if the data set is small. Fracture heights range from 0.1 to 7.60 ft, averaging 1.85 ft (n = 30 fractures).

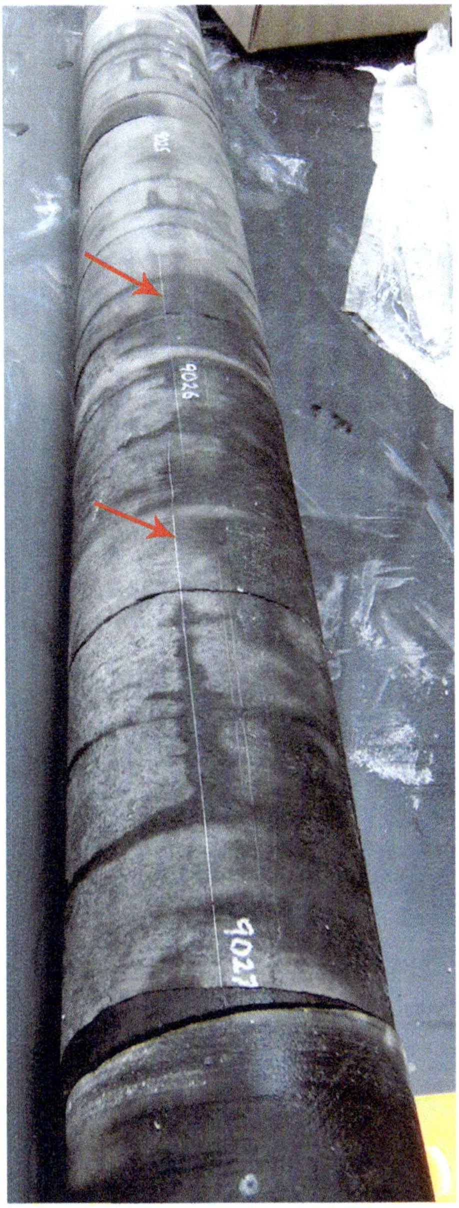

Figure A1c1.3 Two overlapping, narrow, calcite-mineralized, high-angle extension fractures with dips parallel to the core axis so that the full heights can be measured and logged. Vertical four-inch sandstone core, uphole is towards the top of the photo.

that the unstimulated fracture-controlled vertical permeability may also be vertically limited.

- Termination at a stylolite, where the fracture formed after the stylolite which acted as a mechanical boundary to arrest fracture propagation. Alternatively, postfracturing stylolite-related dissolution may have shortened the fracture.
- Blind terminations within an apparently homogeneous lithology, indicating that the strain at the time of fracturing was insufficient to propagate the fracture to a bedding contact.

Figure A1c1.4 Even thin shale layers can provide enough mechanical contrast to arrest the propagation of extension fractures, suggesting that the energy required to form this type of fracture is relatively low. Vertical four-inch diameter limestone core, uphole is towards the top of the photo.

Unknown Terminations

- Out of core, i.e., the fracture exits the side of a core or sometimes out the very top or bottom of a core before terminating.
- Unknown, which includes fracture terminations that cannot be pinpointed but that probably occur within missing core pieces or within rubble zones.

Vertical extension fractures commonly terminate against an adjacent, more ductile lithology (Figures A1c1.4–A1c1.8), where strain that was accommodated in brittle fashion in the fractured layer was accommodated by ductal flow in the non-fractured layer. This type of termination is one of the common although not universally applicable criteria for identifying extension fractures.

Extension fractures may also terminate blindly within an apparently homogeneous lithology (Figures A1c1.9, A1c1.10). In these cases, the stress that was driving fracture propagation was insufficient to extend the fracture to the bed boundary.

Extension fractures also commonly terminate at stylolites. It is not always obvious whether the stylolite predates fracturing and provided a mechanical contrast against which the fracture terminated, or whether

Figure A1c1.5 Mechanical properties can change over geologic time due to diagenesis and changes in pore pressures. Some clay-rich lithologies that are presently relatively ductile were, under different conditions, the more fracture-prone strata in a heterogeneous sequence. In this example, a calcite-mineralized vertical extension fracture in a shale tapers towards its termination against a limestone layer. At the time of fracturing, the shale was more susceptible to fracturing than the limestone. Slab from a vertical three-inch diameter core, uphole is towards the top of the photo.

Figure A1c1.6 A near-vertical, calcite-mineralized fracture exits the core downward before reaching its termination, and terminates upward, with twist hackle, against a lithologic boundary. The upward decrease in the thickness of the calcite mineralization corresponds to a narrowing of the fracture. Vertical four-inch diameter limestone core, uphole is towards the top of the photo.

Figure A1c1.7 Short, calcite-mineralized, strata-bound fractures that are confined to brittle limestone layers and that terminate abruptly against the interbedded ductile shales. Slab from a vertical four-inch diameter core, uphole is towards the top of both photos.

Figure A1c1.8 A strata-bound, calcite-cemented vertical extension fracture in a gray dolomite terminates abruptly, without tapering, at a contact with an adjacent limestone. Vertical four-inch diameter core, uphole is towards the top of the photo.

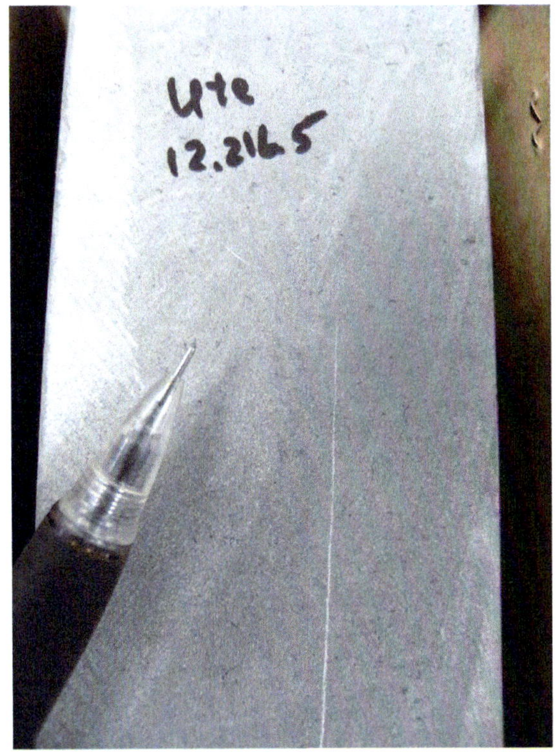

Figure A1c1.9 Some lithologic changes are gradual, as in this upward transition from fine-grained sandstone to siltstone. The gradual narrowing and termination of this calcite-mineralized, high-angle extension fracture reflect this gradual change in lithology. Slab from a vertical four-inch diameter core, uphole is towards the top of the photo.

Figure A1c1.10 An irregular high-angle extension fracture that terminates blindly, i.e., for no mechanical reason, in a relatively homogeneous, cross-bedded, anhydrite-cemented, eolian sandstone. Vertical, 2.5-inch diameter core, uphole is towards the top of the photo.

fracturing was first and the fractures were truncated as the rock dissolved along the stylolite.

Other fractures exit the sides of a core before terminating (Figures A1c1.11–A1c1.13), sampling only a fraction of the total fracture height.

Some extension fractures terminate blindly but overlap with similar, parallel fracture segments (Figures A1c1.14–A1c1.17). In some examples the overlapping fracture tips hook towards each other, suggesting that two fractures propagated towards each other in nearly the same plane. Other fracture pairs show no such interaction, and some are in fact segments of the same fracture, joining together into a single plane in the third dimension. Given the limited volume of a core sample, it can be unclear whether such fracture segments form a single permeability conduit or whether they are separate fractures and the fracture-controlled permeability is discontinuous. If they are single conduits, the en echelon segments should be measured as a single fracture height.

Figure A1c1.11 Left: high-angle, calcite-mineralized extension fractures in limestone terminate at the insoluble clay residue along an irregular, stylolitized shale parting. Right: plan view of a stylolitized shale bedding plane in the same core, showing muddy ridges that project into the dissolved ends of the fractures where they terminate against the shale parting, suggesting that dissolution/stylolite formation postdated fracturing. Fracture-filling calcite is commonly more soluble than a host limestone, and clays along the parting conformed to linear dissolution embayments along the fractures. Vertical four-inch diameter core, uphole is towards the top of the left photo, and away from the viewer in the right photo.

Figure A1c1.12 Full fracture height cannot be measured where the fracture exits the core top or bottom before terminating. In the example on the left, the location and type of termination at the base of a calcite-mineralized extension fracture are unknown since the fracture exits the side of the core downward (the fracture is highlighted by the dashed line marked on the core surface beside the fracture trace). In the right photo it is the upper termination that is unknown. Even though truncated, minimum heights can and should be measured. Vertical, four-inch diameter limestone (left) and shale (right) cores; uphole is toward the tops of both photos. The core on the left was also oriented relative to north, as indicated by the scribe groove immediately to the left of the black line. The red–black line pair in this photo is the most common convention for marking the in situ up orientation of a core, with "red on the right looking uphole."

Figure A1c1.13 One face of a high-angle extension fracture skims the edge of this core. A small but important data set can be recorded from this example, including parameters such as the presence of fracturing, a minimum fracture height and width, and the fact that the fracture is mineralized with calcite but that the mineralization is weaker than the host rock. Fracture strike might also be addressed if this fracture plane can be related to the strike of other induced or natural fractures. Vertical, four-inch diameter limestone core; uphole, as marked by both the red–black line pair and the uphole arrows, is towards the top of the photo.

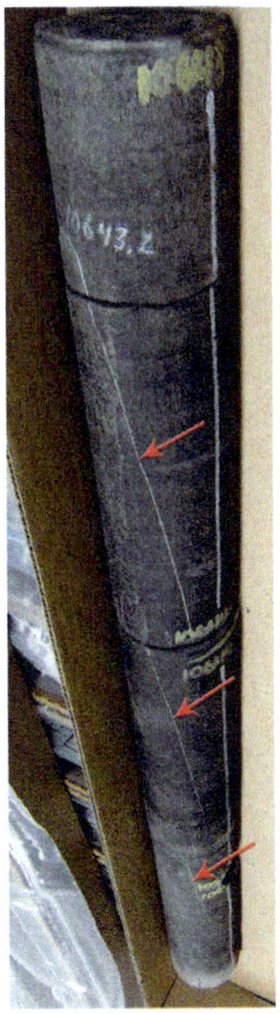

Figure A1c1.14 Three en echelon extension fractures in a calcareous shale. Limited vertical fracture-controlled permeability is suggested by fractures that overlap but do not join in the third dimension, at least not within the volume of the core. Vertical, three-inch diameter core; top of the core is towards the top of the photo.

Figure A1c1.15 En echelon segments that hook towards each other suggest separate fractures that propagated towards each other. When traced across the end of the core, these two fractures consist of separate, parallel planes. Vertical, three-inch diameter shale core; uphole is towards the top of the photo.

Figure A1c1.16 Some overlapping en echelon offsets occur at mechanical discontinuities such as variations in lithology. Here, the propagation of a fracture in shale was inhibited by thin limestone beds, but equal strain across the formation created a new fracture that propagated in a parallel, offset plane. Slabs of vertical, four-inch core, uphole is towards the top of the photo.

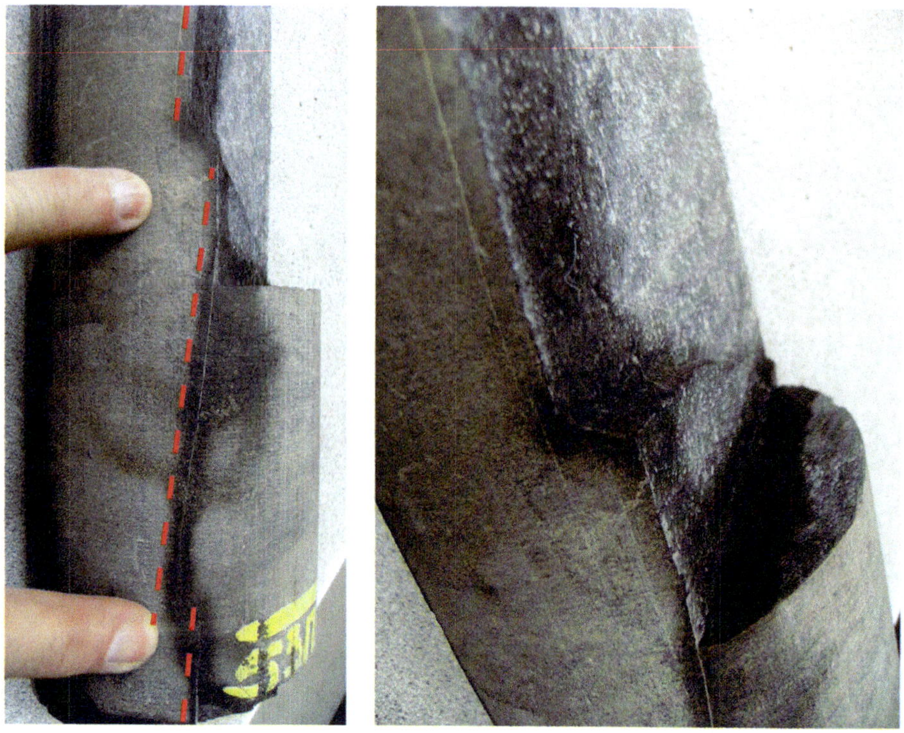

Figure A1c1.17 Two views of an en echelon fracture pattern that is only the surficial expression of a single fracture. Left: the surface of the core shows two offsets (at the fingertips) of a calcite-mineralized, high-angle extension fracture system that might suggest three separate fractures with three measurable heights, and discontinuous vertical fracture-controlled permeability. Right: same fracture system viewed oblique to the fracture face, showing that the fracture planes that appear to be separate on the core surface join in the third dimension. This suggests that the en echelon pattern may be twist hackle at the edge of the larger extension fracture. En echelon patterns can also form in shear, and the full 3D geometry of the fracture should be investigated before making an interpretation. Vertical, four-inch diameter limestone core; top of the core is towards the top of the photos.

A1c2 Extension Fracture Lengths and Lateral Terminations

Except for the smallest fractures (Figures A1c2.1–A1c2.4), horizontal fracture lengths cannot be assessed with vertical core. However, where extension fracture lengths can be measured in outcrop (e.g., Lorenz and Laubach, 1994), they commonly have lognormal distributions similar to other fracture parameters. Unfortunately, the dimensions of outcrops that expose bedding planes and that therefore might be useful for

Figure A1c2.1 Short, limited-length, poorly mineralized, high-angle extension fractures in a shaley limestone. Left: Oil bled from the fractures onto the surface of the core. Right: The fractures are only a few centimeters long and a few centimeters tall. Numerically this zone forms a spike in plots showing fracture frequency by depth, but the fractures are parallel and poorly interconnected. Vertical, four-inch diameter core; uphole is towards the top of the photo on the left, and the view is downhole on the right.

Figure A1c2.2 A circular, calcite-mineralized extension fracture in a poorly bedded shale. The fracture is only about 0.1 mm wide, but its prominent intersection with the slab face inaccurately suggests that it is an important contributor to permeability. It is part of a suite of scattered similar fractures in the shale core, terminating blindly top and bottom rather than at bedding planes. Limited observations suggest that the fractures of this suite are parallel and nearly circular. The photo on the left displays the opposing faces of the fracture, covered with drilling mud. The photo on the right shows one of the fracture faces after the mud has been washed off. It is still wet, obscuring the thin layer of calcite cement. Slabs of four-inch vertical core; uphole is towards the top of both photos.

Figure A1c2.3 Opposing faces of a small, circular, calcite-mineralized, high-angle extension fracture, exposed in butterfly fashion. Secondary induced fracturing has extended the fracture plane beyond the mineralized natural fracture edges, into freshly broken rock (lighter gray areas), along a plane that is slightly oblique to the original mineralized fracture. Four-inch diameter vertical shale core; uphole is towards the top of the photo.

Figure A1c2.4 A calcite-mineralized high angle extension fracture face in shale. Mineralization, terminating along an arcuate front within the core, defines the lateral limit of the fracture prior to breakage of the core. The thumb is resting on a second, parallel, and slightly offset mineralized fracture plane. The two fracture planes are separated laterally by a slab of rock a few millimeters thick. Vertical three-inch core; uphole is towards the top of the photo.

measuring fracture lengths are commonly smaller than fracture lengths. Few outcrops provide large enough areas to capture one or both of the lateral terminations of a significant number of fractures, and there are few published data sets from such pavements. Where fracturing is well developed, the tips of closely spaced extension fractures may overlap and, as they do in vertical exposures, the two fracture segments may merge in the third dimension. Extension fracture lengths may be effectively unlimited where a system of parallel extension fractures is well developed, creating elliptical and highly anisotropic reservoir drainage.

High-angle extension fractures that are not strata-bound are commonly circular or slightly oval in shape when viewed normal to the fracture face. However, once their heights reach the mechanical bedding-plane limit, extension fractures grow primarily by extending parallel to bedding and fracture lengths may be significantly larger than fracture heights. Since length is governed by strain whereas height is controlled by bedding thickness, cross-plots show no relationship between the two parameters.

The lateral terminations of extension fractures are occasionally found in outcrops and less frequently in core. Lateral terminations can be recognized by an arcuate fracture front and locally by a limit of mineralization. Sometimes the front separates the natural fracture surface from a rougher, unmineralized, induced fracture that extends outward from the natural fracture. The lateral edges of high-angle extension fractures can also be marked by twist hackle.

A1c3 Extension Fracture Spacings

Lateral fracture spacing is one of the more important yet also one of the more difficult parameters to obtain when trying to assess fracture contributions to reservoir porosity and permeability. As with the length and width, extension fracture spacing populations commonly have lognormal distributions that consist of numerous closely spaced fractures and increasingly smaller populations of wider spacings (Figures A1c3.1–A1c3.2). Vertical core is a poor mechanism for sampling the lateral spacings of high-angle fractures (Figure A1c3.3), but cores can and do commonly capture spacings that are less than the core diameter, indicating that fracture spacing can be quite close. Such spacing measurements typically represent only the low end of a lognormal spacing distribution.

Several examples of closely spaced mineralized extension fractures within a core diameter are illustrated. The available spacing measurements can be treated statistically (e.g., minimum, maximum, and average) but in fact, these statistics may not be representative of the total spacing population since spacings wider than core diameter are not included.

Narr (1996) offered a method for obtaining a number for the lateral spacing of high-angle fractures from vertical core data, with certain constraints. Narr's two deceptively simple formulae for fracture spacing are:

- Spacing = (average aperture × core diameter × ft of fracture-prone core), divided by (the sum of apertures × sum of fracture heights)
- Spacing = (core diameter × ft of fracture-prone core), divided by (the sum of fracture heights).

The required conditions for the application of these formulae are that:

- a single set of parallel fractures is being measured (as in Figures A1c3.4, A1c3.5)
- all fractures are oriented normal to bedding and bedding is oriented normal to the core axis
- fracture height is significantly greater than the core diameter
- a representative sample of the fracture population has been captured by the core
- fracture apertures are consistent

A spacing calculated from these formulae is a single number and does not define the more common range of spacings. Nevertheless, if the conditions are met, Narr's formulae can provide a spacing value where none existed.

If the lateral fracture spacing cannot be calculated from vertical core, spacing estimates fall back on experience in comparing fracture intensities in vertical wells to data from horizontal wells in the same formation (e.g., Lorenz and Hill, 1994). In general, we have found that a formation probably contains enough fractures to significantly affect a reservoir if *any* vertical fractures are captured by 10 ft of vertical core. Conversely, the formation *cannot* be assumed to be unfractured if there are no fractures in the core.

Experience also suggests that 1 foot of cumulative vertical fracture height per 10 feet of vertical four-inch diameter core marks the minimum ratio for a significant fracture intensity in many reservoirs. This proxy for fracture development commonly translates into an average spacing of about 3 feet, but that average occurs within a range of a few inches to a few tens of feet.

Cores from deviated wells (Figures A1c3.6, A1c3.7) do a much better job of sampling the lateral spacing of high-angle fractures, but even those wells must have azimuths that cut across or at least oblique rather than parallel to the fracture strikes in order to acquire a statistically valid sample of the spacing population. Raw fracture spacing measurements along the axis of a horizontal core should be geometrically corrected (the "Terzaghi correction") to get true fracture spacings normal to the fracture planes. If this is not done, closely spaced fractures that strike nearly parallel to the axis of the core may appear to be less well developed than a set of more widely spaced fractures that strike normal to the core axis.

Core from inclined wells can also offer reliable fracture spacing measurements where two or more parallel fractures are present in the same piece of core. Where there are few mechanical bedding discontinuities and the fractures can therefore be extrapolated vertically with a degree of confidence, reasonable spacing measurements can be obtained by measuring the distance between fractures along the core, making it the hypotenuse of a triangle, and then geometrically calculating the distance between the two fractures normal to the fracture planes.

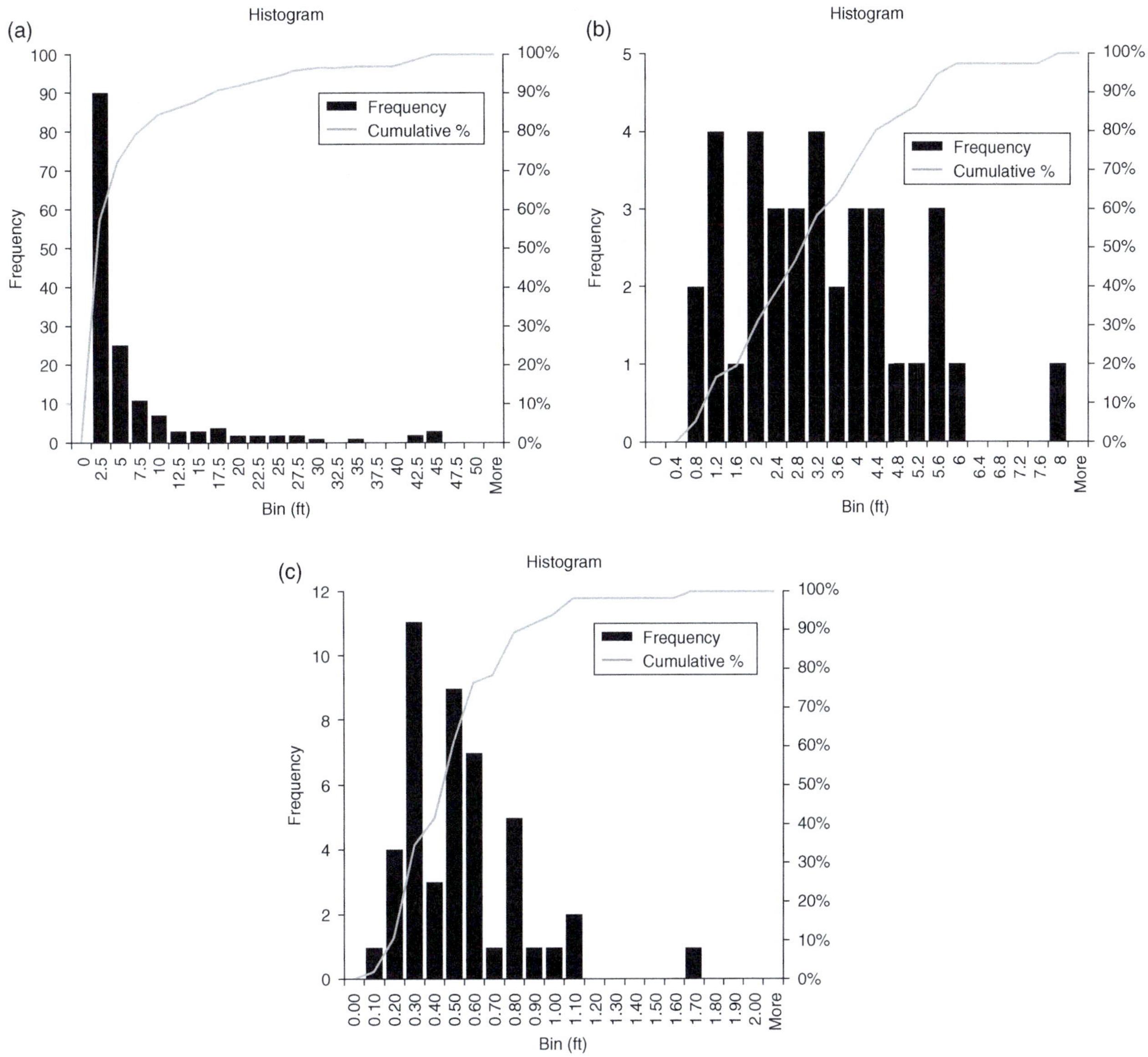

Figure A1c3.1 Histograms of the spacings of high-angle extension fractures in horizontal cores cut from three different formations. Upper left: data from a four-inch core cut from a calcareous, deep marine shale (n = 158). Spacings in this core range from 0.01 to 43.77 ft and average 5.54 ft, histogram bin size 2.5ft. Upper right: data from a four-inch core cut from a deep marine, fine-grained sandstone (n = 36). Spacings range from 0.46 to 7.64 ft, averaging 3.06 ft, histogram bin size 0.4 ft. Bottom: data from a four-inch core cut from a deep marine, clay-rich shale; the plotted fractures are confined to a calcareous, 2.5 inch/6.3 cm thick bed within the shale that was cored parallel to bedding (n = 46). The spacings in this bed range from 0.1 to 1.67 ft and average 0.52 ft, histogram bin size 0.10 ft. The drop-off in frequency to the left that is shown in two of these histograms is common, but can be lost depending on the bin size of the plots.

Figure A1c3.2 Photo showing the spacing patterns for two sets of bed-normal extension fractures on the inclined upper bedding surface (dipping uniformly at about 20° towards the viewer) of a marine sandstone. The fractures extending top to bottom of the photo have a regular distribution whereas the more closely spaced fractures extending across the photo have a lognormal spacing pattern. The fact that there are two distinct spacing patterns in this one sandstone demonstrates that the rule of thumb that spacing is proportional to bed thickness is only useful as a first approximation in the absence of other data; bed thickness is only one of several controls on fracture spacing, and not typically the dominant one.

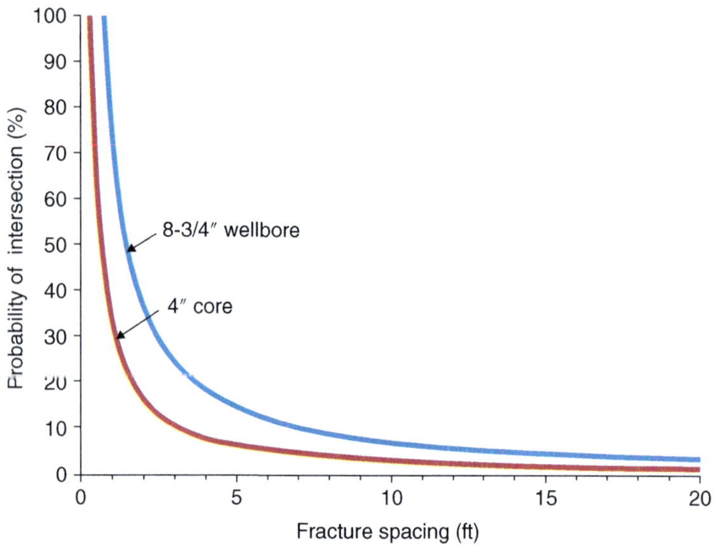

Figure A1c3.3 The probability of intersecting a vertical fracture with a vertical wellbore or a vertical core is low until fractures are closely spaced. A four-inch diameter core has only a 50% chance of capturing a fracture if the fractures have an average eight-inch spacing (from Lorenz, 1992).

Figure A1c3.4 Two views of closely spaced, strata-bound, parallel, unmineralized, high-angle extension fractures. Left: fractures are parallel to the white lines, and are confined to thin siliceous shale beds bounded by thinner clayey shale beds. Right: same piece of core, looking from the top down and showing the two closely spaced fractures in the upper layer of siliceous shale. The presence of this many fractures suggests that despite the small sample size, the core captured a representative population of both spacings and heights, and that the reservoir is intensely fractured. Vertical 3.5-inch core; uphole is towards the top of the photo (left) and towards the viewer (right).

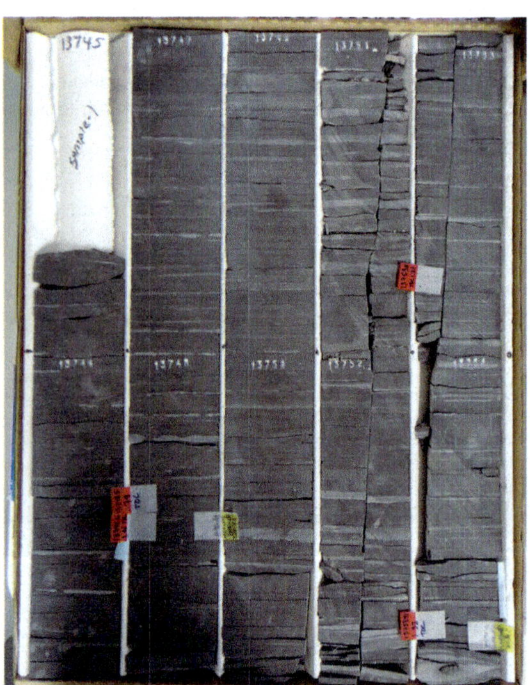

Figure A1c3.5 A near-vertical extension fracture that extends from the base of the middle row of core slabs, through the fourth row, and out of the core in the fifth/last row of slabs on the right. The mineralized fracture is at least four feet (1.3 m) tall and cuts indiscriminately across the poorly developed bedding in this shale formation. A similar fracture is present in the next box of core slabs. The fractures have parallel dip angles and are assumed to have parallel strikes although that cannot be definitively determined since the core cannot be locked together. If they have parallel strikes and extend vertically far enough to overlap, as is probable, then the spacing between them can be calculated using basic geometry, and the two fractures are separated by only 1.4 ft of rock.

Figure A1c3.6 High-angle extension fractures in a near-horizontal core, in a glauconitic limestone near the contact with an underlying shale. Three parallel natural fractures are located at the tips of the blue pencil, the white pen, and the white toothbrush handle (the toothbrush is used for cleaning the core). The flashlight and gray pencil lying parallel to the core were used for stabilizing the core pieces for the photograph. The bed-normal natural fractures are obscured by various saw cuts and induced breaks in the core. If this spacing were typical of the reservoir, a vertical core would have about a 75% chance of capturing one of these three fractures. However, fractures are more widely spaced, of the order of 10 ft, in the middle of the limestone, reducing the probability of fracture intersection to about 3%. Butts of four-inch core; uphole is towards the right of the photo.

Figure A1c3.7 The spacings of high-angle extension fractures can be measured directly in many horizontal cores. The scribe-line groove next to the red line on the core surface indicates that this sandstone core was mechanically oriented relative to stratigraphic up, and the fractures were determined to be vertical, corroborated by their orientation normal to bedding. Bedding planes may be horizontal, but there can still be a 180° ambiguity as to top vs. bottom of the bedding. That ambiguity was resolved for this core with the orientation survey. Fracture spacing normal to the fracture planes, as opposed to their spacings along the core axis, can be measured directly. Note that the "uphole" core orientation line pair has non-standard colors: red is on the right looking uphole as usual, but a white line replaces the more common black line in the pair. In horizontal cores, "uphole" refers to the direction towards the heel of the well and then upward to the surface. Two red–white line pairs were marked on this core so that both the slabs and butts would be marked after slabbing, but there can be confusion if the pairs are drawn too close together. Horizontal four-inch diameter core; uphole is to the left.

A1d Extension Fracture Variations and Lithologic Influences

The ideal extension fracture is planar, with parallel opposing faces and a reasonably uniform width except where it pinches out and terminates. However, the host lithology influences the characteristics of a fracture, and the same stress system that generates planar and systematic fractures in a uniform, fine-grained lithology may form rough and irregular fractures in the interbedded coarse-grained lithologies.

Extension fractures are commonly more intensely developed in the brittle strata in a formation and may even be restricted to those units. For example, sandstones are typically more heavily fractured than interbedded muddy shales, and dolomites are typically more heavily fractured than interbedded limestones. However, this is not a universal relationship; the mechanical properties of a rock are a primary control on fracture intensity, but geologic systems are not static. Mechanical properties change with diagenesis, cementation, compaction, and temperature. Confining stresses and pore pressures change with burial depth, also affecting the mechanical properties of a rock, while fracture development changes with strain magnitude and rate.

The effects of these variables on fracturing can be quantified in laboratory experiments where one or two parameters are changed at a time, but in nature these controlling parameters are poorly constrained. Moreover, in core we are presented merely with the results of several of nature's superimposed experiments. Natural fracturing was controlled by the stress conditions present in the formation and the mechanical properties of the rock at the time of fracturing, not by the present-day stresses and rock properties.

In some sandstone formations, shale partings only a few centimeters thick have arrested the propagation of meter-tall extension fractures. Under other conditions, however, extension fractures, driven by a high in situ stress anisotropy or reacting to low mechanical contrasts in the different layers, crossed bedding boundaries, disregarding the geomechanical differences caused by layering. It is also possible to find well-developed extension fracture systems in thick shale units since elevated pore pressures can change even clay-rich shales, commonly considered to be relatively ductile, into brittle and fracture-prone strata.

Fracturing in a heterogeneous formation may contain entirely different fracture sets in the different layers (Figure A1d.1), all formed under the same stress system, due to varying mechanical properties of the layers (e.g., Lorenz et al., 2002). Elsewhere, unrelated fracture sets have been superimposed on a given layer, reflecting different stress conditions and mechanical properties in the layer at different times in its history. Numerous outcrop examples show multiple, unrelated sets of extension fractures, with different intensities and different strikes, in the same layer. A single layer of coarse-grained sandstone of the Permian Abo sandstone in central New Mexico contains both a set of dip-slip conjugate shear fractures formed under one set of conditions and, striking at right angles to the shear fractures, a set of high-angle extension fractures formed under significantly different conditions.

Many extension fractures are not marked by plume structures. Interpretations of these as extension fractures rely on other features that are common to extension fractures such as terminations at minor bedding contrasts and an absence of evidence for shear.

A1d1 High-Angle Extension Fractures in Limestone

Extension fractures in limestone are as variable as the lithologies that fall into the category of "limestone." Fine-grained micrites commonly host narrow extension fractures marked by plume structure (Figure A1d1.1), but extension fractures are likely to be rougher and more irregular in vuggy (Figure A1d1.2) and coarser-grained (Figure A1d1.3) limestones.

Figure A1d.1 Multiple narrow, calcite-mineralized, high-angle extension fractures in a limestone accommodated the same strain that was taken up by a smaller number of wider fractures in the underlying calcareous shale. (The apparent offset at the horizontal plane below the lithologic contact is due to differences in the orientations of the slab planes above and below the offset, and is not real.) Vertical three-inch diameter core; uphole is towards the top of the photo.

A1d2 High-Angle Extension Fractures in Dolomite

Extension fractures in dolomites are commonly rough and irregular (Figures A1d2.1–A1d2.4) because dolomites tend to be coarse-grained and are commonly vuggy. The fracture faces are rarely marked by plume structure since the the surface relief on a plume is typically less than dolomite grain size. The diagenesis, dissolution, and reprecipitation associated with dolomites typically result in irregular and wide fracture apertures with irregular mineralization.

A1d3 High-Angle Extension Fractures in Shale

High-angle extension fractures in cored shales range from a few centimeters to tens of feet tall. Short fractures are common where the fractures are bounded by strong mechanical contrasts between thin beds of different shaley lithologies, but short fractures can also occur within thicker-bedded shales where minimal strain did not propagate tall fractures to the bedding boundaries. Likewise,

Figure A1d1.1 The plume structure commonly associated with extension fracturing can be formed in many lithologies, including limestone as shown here. Plume structure may be inhibited in coarse-grained lithologies such as conglomerates and crystalline dolomites, but can form easily in these strata if the grains do not offer significant mechanical contrasts or if the stress differential was high enough to drive fractures indiscriminately across the grain boundaries. Vertical, four-inch diameter core; uphole is towards the top of the photo.

Figure A1d1.2 Two rough, parallel, high-angle extension fractures in a vuggy limestone, striking oblique to the slab surface (angling into the plane of the photo towards the left). The oblique strike exaggerates the fracture width and irregularity. Fracture dimensions on a slab plane are apparent dimensions unless the fracture strikes normal to the slab surface. Top and bottom fracture terminations are blind within a homogeneous lithology. Slab from a four-inch diameter core; uphole is towards the top of the photo.

Figure A1d1.3 Two views of a rough-surfaced, high-angle extension fracture (between the two arrows in the left photo) in a vuggy, coarse-grained limestone. The photo on the right shows the fracture face, with subtle crystalline calcite mineralization. Plume structure is unlikely to have developed on such irregular fractures in coarse-grained lithologies, but plumes are also easily removed by dissolution in soluble lithologies. Slab of a four-inch, vertical core; uphole is towards the top of both photos.

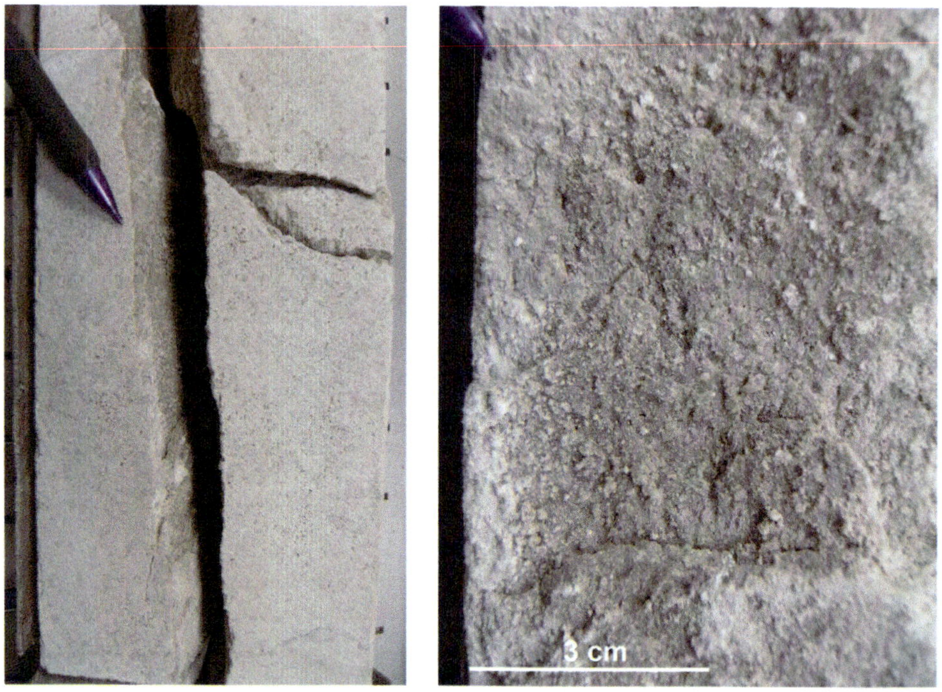

Figure A1d2.1 Two views of an extension fracture in a coarsely crystalline dolomite. Fractures in dolomites tend to be irregular and to have rough surfaces, due to propagation around the dolomite rhombs and the typically multiphase diagenetic history of these rocks. Fracture faces may be sparsely mineralized with isolated knobs and crystals of dolomite (right). The apparent width of the fracture on the left is due only to the way the broken fracture is lying in the box; actual width is on the order of a few millimeters. Vertical three-inch diameter core; uphole is towards the top of both photos.

Figure A1d2.2 An irregularly planar extension fracture in a dolomite. If the fracture is irregular in only one plane, i.e., if horizontal cross-sections across this core were to show linear sections of the fracture at any given depth despite this irregularity in the vertical plane shown here, it would be plausibly interpreted as a strike-slip shear fracture, especially if the surface was marked by shear indicators. However, the fracture is likely to be an extension fracture if the fracture surface is irregular in cross-sections oriented in any direction. This core has broken open along the poorly mineralized fracture plane. Butts of vertical four-inch core; up-section is towards the top of the photo.

Figure A1d2.3 Left: a tall high-angle extension fracture in a coarsely crystalline dolomite has a rough, irregular plane. The core has broken open along the incompletely mineralized fracture plane. The green line between the red–black uphole/downhole orientation lines is a Master Orientation Line useful for assessing fracture strikes relative to each other in continuous intervals of core. Right: scattered millimeter-scale dolomite crystals are present on the faces of a fracture from the same formation and fracture system. Some of the irregularity of this fracture is due to postfracturing dissolution. The dolomite crystals were precipitated after dissolution. Vertical four-inch diameter core: uphole is towards the top of both photos.

extension fractures may be narrow (Figure A1d3.1) or wide (Figure A1d3.2), depending on the amount of strain that each fracture had to accommodate. Extension fractures in shales are commonly marked by plume structure (Figure A1d3.3), although it may be subtle and is easily obscured by mineralization.

Most shale fractures are mineralized to some degree, but recent studies (e.g., Landry et al., 2015) suggest that even mineralized, low-permeability fractures in a shale can have higher permeability than the typically nanodarcy scale permeability of a hosting shale. Fractures in shales provide large surface areas for the diffusion of fluid or gas from the matrix into the fracture, and, hopefully, to a wellbore.

Tall fractures (Figures A1d3.4, A1d3.5) are common in thickly bedded shales where the horizontal mechanical

Figure A1d3.1 Two views of a short, narrow, calcite-mineralized, high-angle extension fracture in a shale core. Left: the fracture (arrow) cuts the slab face. (The arcuate pattern sweeping from upper right to lower left across the slab is scarring from the slab saw.) Right: the two faces of the fracture are mineralized with a layer of calcite that is 0.01 mm thick. The single layer of calcite adheres irregularly to the opposing fracture faces, resulting in what only appear to be unmineralized patches on each face. Butts from a vertical, three-inch diameter core; uphole is towards the top of both photos.

Figure A1d3.2 Two views of a tall high-angle extension fracture in shale. The fracture is approximately 3 mm wide. The photo on the right shows the face of the fracture on the piece of core indicated by the arrow in the photo on the left. In contrast to the fracture shown in the previous photo, both fracture faces are incompletely mineralized with calcite that grew from the walls into the open aperture. Irregular mineralization indicates that this fracture creates an important conduit to fluid flow, with permeability significantly greater than that of the matrix of the host rock. Slabs of vertical, four-inch diameter core; uphole is towards the top of the photo.

Figure A1d3.3 Plume structure, recording horizontal propagation of a short, strata-bound, high-angle extension fracture follows bedding in a thinly bedded shale where the different layers provide strong mechanical contrasts. Fracturing is confined to the more siliceous beds in the formation. Vertical three-inch diameter core; uphole is towards the top of the photo.

bedding contrasts are both minimal and far apart. Short fractures are typical of heterogeneous shale formations where they may be confined to the shaley beds (Figure A1d3.6) or to the more calcareous or more siliceous beds (Figure A1d3.7). Some shale formations contain two fracture sets, one in the muddier beds and another with a different strike in the more brittle lithologies. Tall extension fractures can also cut across bedding and multiple lithologies if the driving stress anisotropy was high and/or the mechanical contrast between beds was low (Figure A1d3.8).

Figure A1d3.4 A mineralized, 4 mm wide, high-angle extension fracture in poorly bedded shale extends for several feet nearly parallel to the core axis. The fracture walls are smooth and fracture width is uniform, but the aperture remaining between crystalline calcite that lines the wall is less regular. Calcite mineralization adheres weakly to the fracture walls so the core breaks easily along the fracture plane. Gaps that show in the mineralization are areas where the layer of calcite has become detached from both walls and is missing. The mineralization is the same color and has the same texture as the styrofoam bed on which the slabs are lying. Slabs of vertical, four-inch diameter core; uphole is towards the top of the photo.

Figure A1d3.5 A tall, calcite-mineralized fracture in bedded shale can be traced for about four feet in core slabs. The offset in the fracture at the arrow is spurious, due only to a change in the slab-plane orientation. Reassembly of the slabs and butts shows that the fracture consists of a single continuous plane that has about 50% more height than is apparent in the slabs shown here, and that it exits the sides of the core before terminating so that its full height is greater still. Calcite mineralization is of the order of 0.2 mm thick. Slabs of four-inch vertical core; uphole is towards the top of the photo.

Figure A1d3.6 Two incompletely mineralized, high-angle extension fractures that are confined to shale units interbedded with shell-hash limestones. At the time of fracturing, the shale was more susceptible to brittle deformation than the limestone. Slabs of vertical, four-inch diameter core; uphole is towards the top of the photo.

Figure A1d3.7 A calcite-mineralized, vertical extension fracture that is confined to a limestone interbedded with shale. This fracture extends a short distance into the confining shales, but most such fractures terminate at the bedding contact. In contrast to the previous example, the limestone was more susceptible to fracturing than the adjacent shale. Slabs of vertical, four-inch diameter core; uphole is towards the top of the photo.

Figure A1d3.8 A relatively wide (2 mm), incompletely mineralized, near-vertical extension fracture that extends indiscriminately across shale and limestone beds. The fracture surfaces are slightly rougher in the limestone but otherwise there is little difference in fracture characteristics in the two lithologies. Mineralization consists of two layers of calcite attached to the fracture faces and growing inward, with euhedral crystal faces marking a medial, unmineralized aperture. Slabs of vertical, 3.5-inch diameter core; uphole is towards the top of the photo.

A1d4 Narrow Extension Fractures

Some high-angle extension fracture sets consist of very narrow (0.01 mm or less), parallel to subparallel, mineralized fractures (Figures A1d4.1–A1d4.3). Such narrow fractures can occur either singly or in bundles, and are most common in fine-grained, micritic limestones and homogeneous calcareous shales although they have also been found in sandstones (Figure A1d4.4). These very narrow fractures can be too small and too tightly cemented to create significant mechanical discontinuities in the rock, particularly where they are filled with finely crystalline calcite that has geomechanical properties that are similar to those of the limestone host rock. Younger fractures, including induced fractures, can cut at shallow angles across such fractures with no apparent interaction.

When a rock is strained to the point of fracture multiple times, fractures may repeatedly open and heal with mineralization ("crack-seal"), or they may open incrementally without concurrent mineralization. Narrow fractures may also form once and heal with mineralization that is as strong as or stronger than the matrix rock,

Figure A1d4.1 Five narrow, near-parallel, calcite-mineralized, high-angle extension fractures in a dense chalk cut from an inclined wellbore. The three middle fractures are slightly oblique to each other and cut across one another without apparent interference, suggesting the mineralized fracture plane had mechanical properties nearly identical to those of the host rock. The groove down the center of the photo is a core-orientation scribe line. The core is $2^{3/8}$ inches in diameter and was cut from a wellbore that is inclined about 45° from the vertical. Uphole is towards the top of the photo, and the bedding contact below the depth marking defines horizontal.

Figure A1d4.2 Two sets of narrow, calcite-mineralized, strata-bound fractures in a narrow limestone band interbedded in a shale formation, cut from inclined strata in a vertical hole. Top: some fractures are inclined relative to the core axis but are normal to the tilted bedding and probably formed prior to tilting. Other fractures are vertical, parallel to the core axis, and oblique to bedding, and probably formed after tilting. Bottom: the same fracture system in core cut a few feet deeper in the hole, showing two strikes for the superimposed, narrow, calcite-mineralized fractures (next to the dotted lines). If the bedding dip azimuth ("dip az") is known, as from an image log, the actual fracture strikes can be reconstructed. The four-inch core is vertical but bedding is inclined; the top photo shows the slabbed surface on the core butts, uphole is towards the upper right corner of the photo. The bottom photo shows the end of the slab and the view is downhole.

Figure A1d4.3 Two sets of narrow, intersecting, calcite-mineralized, high-angle extension fractures exposed on the end of a core cut from a calcareous shale. The single fracture cutting upper right to lower left predates the other fractures since they terminate against it. Examination of the slab face alone, or even of the slab in three dimensions, would not have revealed the second fracture set; the second fracture set is apparent only in the butt section of the core. Butts of a vertical, four-inch diameter core; uphole is towards the viewer.

so that each successive stress event creates new, subparallel fractures rather than reopening earlier fractures. This is especially common in low-porosity, fine-grained carbonates where the relatively soluble matrix provides ready material for rapid mineralization of fractures.

Narrow fractures in clay- or organic-rich micrites may be bundled, with successive narrow fractures forming immediately adjacent and parallel to each other rather than reopening the initial fracture. Where postfracturing dissolution has occurred, dissolution is concentrated along the fracture bundles if the calcite mineralization is more soluble than the host rock.

A1d5 Irregular High-Angle Extension Fractures

Some high-angle extension fractures are quite irregular, often as a result of the interaction between the propagating fracture and significant lithologic heterogeneity (Figures A1d5.1, A1d5.2), although heterogeneous lithologies may also contain quite planar fractures (Figure A1d5.3). Irregular fracturing may also occur within a setting where high pore pressure has created a low stress differential so that the fracture strikes are

Figure A1d4.4 Bundled, narrow, high-angle extension fractures in a deeply buried sandstone. Thin sections (location marked by the yellow rectangle) show fractures both cutting through grains and following grain margins. Slab face of a vertical four-inch diameter core; uphole is towards the top of the photo.

Figure A1d5.1 Some extension fractures found in coarse-grained rock such as in this bioclastic limestone are irregular, the fracture planes having propagated around rather than through the clasts. Vertical three-inch diameter core; uphole is to the top of the photo.

Figure A1d5.2 A calcite-mineralized fracture cuts irregularly through a thinly laminated silt-shale lithology. Poor planarity is a function of lithologic heterogeneity probably combined with a low stress differential during fracturing. Vertical, two-inch diameter core; uphole is towards the top of the photo.

Figure A1d5.3 Extension fractures can be planar even in coarse-grained rock such as this pea-gravel conglomerate, if the rock was well cemented, if the clasts had mechanical properties that are similar to those of the matrix material, and/or if the stress differential that drove fracturing was high. Vertical four-inch diameter core; uphole is towards the viewer.

Figure A1d5.4 High-angle extension fractures can also be irregular in structurally complex settings. This oil-stained extension fracture in a limestone is from a faulted anticline. Complex fractures such as this may be related to local structural complications. Vertical, four-inch diameter core; uphole is towards the top of the photo.

Figure A1d5.5 Irregular, oil-filled, high-angle extension fractures in a bedded limestone. This fracture pattern begins to resemble the non-systematic fracture system hypothesized by Gretener and Feng (1985) for fracturing under conditions of high pore pressure that cause the stress differential to be very low and fracture strikes to be poorly constrained. This type of fracture system is rare. Many fracture systems have been suggested to have been caused by pore pressure that exceeds the minimum in situ compressive stress, but the dependent relationship between pore pressure and effective stress does not support such a mechanism for systematic fracture sets. Vertical four-inch diameter core; uphole is towards the top of the photo.

poorly constrained, or in structurally complex areas (Figures A1d5.4, A1d5.5) where the strata have been subjected to multiple stress events. Since these fractures are poorly planar and have no obvious fractographic markings, their interpretation as extension fractures rests on a lack of evidence for shear offset, and on terminations at minor mechanical discontinuities.

A1e High-Angle Extension Fracture Intersections

When two or more intersecting extension fracture sets occur in a reservoir, fracture-related permeability enhancement may be greatest along the best-developed set (which is not necessarily the oldest set), along the least mineralized set (which is not necessarily the best-developed set), or along the set that is parallel to the maximum compressive stress. Thus it is important to characterize natural fractures and to measure their strikes both relative to each other and relative to stress-controlled induced fractures. Complete fracture characterizations and ultimately an understanding of the fracture–stress interactions must be developed from the limited data provided by a core in order to understand a fracture-controlled permeability system.

Although core is a miniscule sampling of a reservoir, examples of intersecting fracture sets in a core are not uncommon. However, fracture intersections are easily missed if only 2D core photos or core-slab surfaces are studied; none of the following examples of fracture intersections come from exposures on slab surfaces since a full 3D understanding of fractures is required in order to measure relative fracture strikes. It is important to use all the core, butts and slabs together, during a core fracture study, and to examine all surfaces of the core. This increases the sample size and significantly improves the probability of finding all of the fracture sets captured by the core, and of measuring their relative strikes. The use of core slabs only for a fracture study is akin to using only 30% of an image log.

Fracture systems with multiple strikes can be created by strike-slip conjugate shear pairs formed during a single stress event, but they can also be the result of superimposed extension-fracture sets. It is important to interpret the origin of intersecting fracture geometries correctly since conjugate shears and superimposed extension fractures have different strikes relative to the in situ stresses, resulting in either fracture closure or shear during production, with different effects on fracture permeabilities. An advantage of core over image logs is that it is usually easier to tell the difference between shear and extension fractures in a core where the fracture face can often be examined and where fracture details can be observed in three dimensions.

Where intersecting fractures consist of superimposed extension fractures but the fractures cannot be oriented, the two sets may be distinguishable by differences in characteristics such as heights, host lithologies, and type/color/completeness of mineralization. However, the fractures of some unrelated, intersecting sets have similar characteristics, and can therefore only be distinguished by strike. Piecing long core sections together on a work table and comparing strikes is often the only way to determine whether a suite of natural fractures in a core consists of one or two fracture sets.

A1e1 Obvious Intersections

Fractures that intersect within a piece of core are not uncommon (Figures A1e1.1, A1e1.2). The butt ends of cores are commonly the best places to look for and

Figure A1e1.1 Two similar, narrow, calcite-mineralized high-angle extension fractures in a muddy shale core, intersecting at a 70° angle. Other than strike, the only obvious distinction between the two fractures is that one dips nearly parallel to the core axis and the other has an 80° dip. The formation is heavily fractured and other intersections occur in the core, but the two fractures are nearly identical. Vertical four-inch diameter core; uphole is towards the top of the photo. The wider white marking on the left side of the core is a service company orientation mark.

Figure A1e1.2 Intersecting, calcite-mineralized, high-angle extension fractures exposed on the end of a vertical shale core, same core and formation as the previous photo. The core is oriented so we know its in situ position, as marked by the "N" arrow, and we can reconstruct the true fracture strikes, 40–220° (two fractures) and 110–290° (one fracture). The fractures appear to be mutually cross-cutting but close examination of the mineralization shows that one cuts across the other and is younger (see the next photo). Vertical four-inch diameter core; view is downhole.

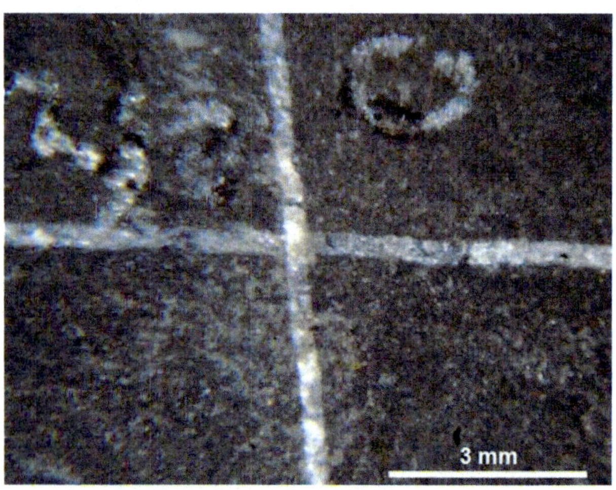

Figure A1e1.3 Close-up of intersecting, calcite-mineralized, vertical extension fractures on the end of a vertical shale core (from the same core shown in the previous photo) showing that the mineralization of the fracture oriented top to bottom in the photo cuts across and postdates the mineralization of the left-to-right fracture. Vertical four-inch diameter core; view is downhole.

document intersections (Figures A1e1.2–A1e1.5). Intersecting fractures may have similar or distinctly different characteristics such as aperture and mineralization (Figure A1e1.6).

Figure A1e1.4 Intersecting, unmineralized, closely spaced, high-angle extension fractures on the end of a vertical limestone core, with local oil stain. The older fracture set is planar and through-going, and has consistent strikes. Younger, oblique fractures are less planar and terminate or are offset at the through-going fractures. Abutting relationships indicate that the fracture set cutting across the photo top to bottom is younger than the left-to-right fracture set. The top-to-bottom fracture set might be interpreted to have been offset by shear along the left-to-right set, but neither the magnitudes nor the senses of offset are consistent, and none of the fracture faces are marked by shear indicators. The top-to-bottom set propagated to and terminated against discontinuities provided by the earlier left-to-right set. Vertical four-inch diameter core. The circled dot indicates that the view is downhole.

Figure A1e1.5 Four calcite-mineralized but incompletely filled fractures with at least three distinct strikes in core cut from a limestone. The relative ages of the fractures are unclear. Water has wicked into the aperture of the lower left fracture indicating significant remnant fracture porosity. Vertical, four-inch diameter core. The circled X indicates that the view is uphole.

A1e2 Less Obvious Intersections

Only one set of a pair of intersecting fractures may be exposed on a slab face (Figure A1e2.1), and intersections may not be obvious from the two-dimensional views provided by photographs (Figure A1e2.2). Incomplete

Figure A1e1.6 Intersecting high-angle extension fractures ("NF") as exposed on the end and slab face of a vertical core. The larger fracture has an irregular width and aperture, suggesting dissolution and indicating good potential for fluid flow. Dissolution-enhanced fractures in this core are oriented at a characteristic angle to the narrower fractures, and the two sets can be distinguished even where they do not intersect. The two planes of exposure, on the slab surface and on the end of the core, illustrate the difference between the true and apparent widths of the larger fracture. Slab from a vertical four-inch diameter core; uphole is towards the top of the photo.

Figure A1e2.1 Fragments of two high-angle extension fractures (planes marked A and B) in the butts of a core cut from dolomite. Only one of the fractures intersects the slab plane. Vertical three-inch diameter core; uphole is towards the top of the photo and away from the viewer.

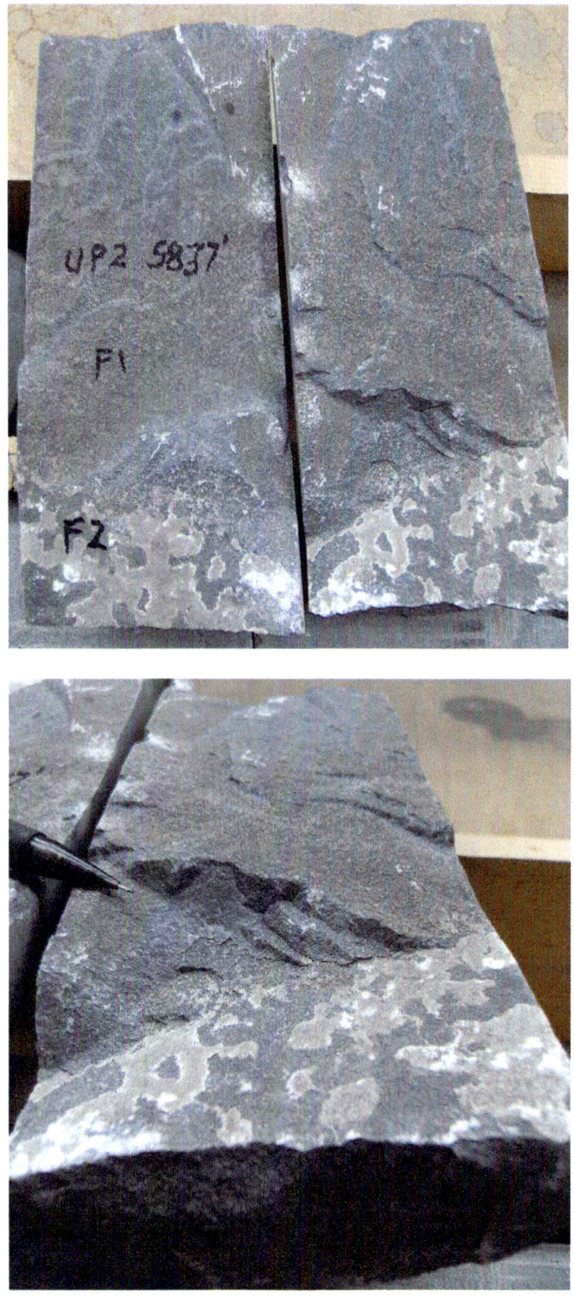

Figure A1e2.2 Two views of fractures in a limestone core. Top: the core is split along a mineralized fracture plane with both faces exposed, showing two types of mineralization. A dirty gray calcite with a crystalline habit covers the upper 80% of the fracture face and shows that significant in situ permeability exists along the fracture. White, amorphous calcite partially occludes the lower part of the fracture. Not apparent from this photo is the fact that the different types of mineralization occur on different, oblique fracture planes, as indicated by the end-on view (bottom) showing that this surface is composed of two fractures with non-parallel strikes. Note that only about 50% of the opposing faces of the lower fracture are covered with the white mineralization, but that the patches of amorphous calcite are mirror images of each other and that the fracture is in fact almost fully occluded. Vertical four-inch diameter core; uphole is towards the top of both photos.

fracture characterizations may be made using core photos, but their limitations must be recognized. Some fracture intersections, and their relative ages, are apparent from steps where one fracture intersected a previous fracture (Figure A1e2.3).

A1e3 Projected Intersections

Fractures with different strikes may not intersect within a core volume, but can easily be projected to an inferred intersection in the nearby rock (Figures A1e3.1, A1e3.2).

Figure A1e2.3 A faint plume on the surface of a vertical extension fracture (parallel to the plane of the photograph) did not propagate cleanly across the oblique, older natural fracture that it intersected (dark, oblique plane along the middle of the core). In some places the younger, plumed fracture has stepped laterally in order to cross this pre-existing mechanical inhomogeneity in the rock, which strikes inward and to the left from the plane of the photo. Vertical four-inch diameter core; uphole is towards the top of the photo.

Figure A1e3.1 Two calcite-mineralized vertical extension fractures skim the edge of a limestone core. The fractures intersect at an 80° angle but they have similar characteristics, and the two sets are indistinguishable where they do not occur together in this unoriented core. The importance of recognizing the presence of one versus two fracture sets is that two sets form an interconnected permeability network whereas a single extension fracture set forms a poorly connected network with highly anisotropic drainage and permeability. Vertical 4.5-inch diameter core; uphole is towards the top of the photo.

Figure A1e3.2 Fracture planes in horizontal cores can also be extended beyond the core volume to projected intersections. Here two sets of high-angle extension fractures in a horizontal core have strikes that intersect at nearly 90°, but the intersections are not captured within the core. The silver line marks the high or dorsal side of the core. A quick, conceptual image of the reservoir fracture system, consisting of a network of closely spaced, high-angle fractures with 90° intersections, can be developed from this core. Horizontal three-inch diameter core; uphole is to the left of the photograph.

Figure A1f.1 Left: edge-on view of two vertical, calcite-mineralized extension fractures (arrows) at the left and right edges of this piece of slabbed sandstone core cut from a deviated well. The wellbore deviation angle was 60° from vertical (wellbore deviation is typically expressed in degrees from vertical, unlike geologic dip angles which are given in degrees from horizontal), and the core is pictured in its in situ position. The unoriented core can be rotated around its axis (*red dashed line*) but the fractures, known to be vertical from nearby vertical cores, are only vertical in one rotational position. The wellbore deviation azimuth is approximately north, therefore these are approximately east-west striking fractures. Right: the face of one of the vertical, calcite-mineralized fractures on the end of the core piece shown in the left-hand photo; calcite partially obscures a plume structure. Deviated, four-inch diameter core: uphole is towards the upper left of the left photo, and away from the viewer in the right photo; stratigraphic up is towards the top of both photos.

A1f High-Angle Extension Fractures in Deviated Core

In contrast to vertical cores, cores from deviated wellbores usually cut across rather than along individual high-angle extension fractures. Deviated cores therefore typically provide smaller samples of individual fractures but have the potential to sample a larger number of fractures. Fracture frequency must be assessed carefully since deviated cores will capture numerous samples of the fractures that strike normal or nearly normal to the wellbore deviation azimuth yet may sample few or none of the fractures striking nearly parallel to that azimuth even if they are more closely spaced.

A deviated core typically provides more and better information on fracture spacings (Figures A1f.1, A1f.2) than vertical cores. On the other hand, a deviated core is less likely to capture the vertical terminations of high-angle extension fractures (Figures A1f.3, A1f.4).

Deviated core commonly affords an opportunity to determine fracture strikes without orienting the core, provided that stratigraphic up in the core can be determined and the wellbore deviation azimuth is known. However, stratigraphic up is no longer equivalent to the obvious "uphole" direction as it is in vertical core, so stratigraphic up in core from a deviated well is not always obvious.

Figure A1f.2 Three obscure, closely spaced, calcite-mineralized vertical extension fractures (between the pairs of black arrows) in a horizontal sandstone core. The groove across the center of the photograph is a scribe line created by the core orientation shoe. The service company used a non-standard red–white line pair for uphole annotation, but the red line is still "red on the right looking uphole" so the heel of the well is towards the left in this photo. Fracture strikes can be measured if the core high side is known and if the wellbore deviation azimuth is known from the wellbore deviation survey. The ridges on the surface of the core are created by the rotating core bit and are normal to bedding. Horizontal, four-inch diameter core; uphole is towards the left, and stratigraphic up is out the side of the core, towards the viewer.

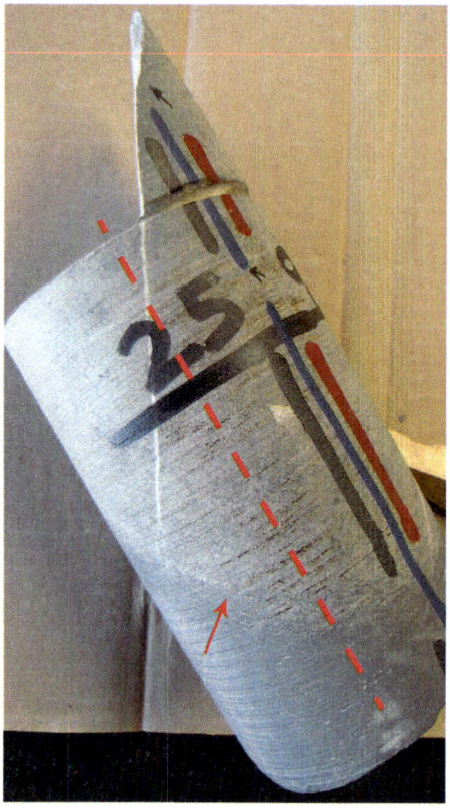

Figure A1f.3 A calcite-mineralized vertical fracture in a piece of unslabbed sandstone core cut from a wellbore that is deviated at about 50° from the vertical. The core has been propped up into its in situ position. There is only one position of rotation around the central core axis (indicated by the red dashed line) at which the bedding (arrow) is horizontal and the fracture is vertical. Fracture strike can then be physically measured using the wellbore deviation azimuth. If bedding is absent or not horizontal, and/or if the fractures are not vertical, the problem becomes more complicated. Deviated, four-inch diameter core; uphole is towards the upper left corner of the photo, stratigraphic up is towards the top of the photo.

Figure A1f.4 Face-on (top) and oblique (bottom) views of a vertical extension fracture that cuts across the axis of a horizontal core. The fracture face above the dotted line has an aged appearance and is more planar than the conchoidal, fresh surface below the dotted line. The upper part of the face is a natural fracture that has been subjected to dissolution; it terminates downward at a bedding plane marked by the dotted line. When the core broke during processing and handling, the natural fracture plane was extended downward as an induced fracture. Horizontal, four-inch diameter core; view is uphole, towards the heel of the deviated well, and stratigraphic up is towards the top of both photos.

Whereas fracture dips can easily be measured in a vertical core, both the dip and strike of a fracture in a deviated core may be unconstrained. Nevertheless, if bedding is arguably horizontal in the formation and can be recognized in the core, there are only two rotational positions around the long axis of the horizontal core where bedding will be horizontal, limiting the number of possible fracture orientations to two.

If the core is inclined relative to bedding and one can tell whether the core is cutting up-section or down-section, then core orientation relative to up can be determined. Moreover, the wellbore deviation survey, which provides an azimuth for the wellbore through the cored interval as well as the deviation of that axis from vertical, can then be used to determine fracture strikes and dips.

For example, if a horizontal wellbore has an azimuth of N60°E and the cored strata are horizontally bedded, a bed-normal fracture that strikes at right angles to the core axis is vertical and has a NNW-SSE strike. If the top of bedding is ambiguous and the fracture is oriented at a 60° angle to the core axis, however, the bed-normal fracture can be reconstructed to have either a NNE-SSW strike or an E-W strike, depending on which side of the horizontal bedding is up. If bedding is not horizontal or cannot be recognized in the core, then the determination of fracture dip and strike in core cut from a deviated wellbore requires that the core be oriented.

When a horizontal core is oriented, the position of the principle scribe line is given in reference to vertical "up", rather than to north as in a vertical core. In the deviated core orientation report, the position of the principle scribe line on the core is given in degrees clockwise from the high side of the core while looking downhole.

References

Gretener, P. E., and Z-M Feng, 1985, Three decades of geopressures insights and enigmas. Bulletin Vereinigung Schweizerischer, Petroleum Geologen und Ingenieure, 51, 1–34.

Landry, C.J., Eichhubl, P., Prodanović, M., and Tokan-Lawal, A., 2015, Permeability of calcite-cemented fractures: flow highway or barrier? AAPG Annual Meeting.

Lorenz, J.C., 1992, Well-bore geometries for optimum fracture characterization and drainage. West Texas Geological Society Bulletin, 32, 5–8.

Lorenz, J.C., and Hill, R.E., 1994, Subsurface fracture spacing: comparison of inferences from slant/horizontal and vertical cores. SPE Formation Evaluation, 9, 66–72.

Lorenz, J.C., and Laubach, S.E., 1994, Description and Interpretation of Natural Fracture Patterns in Sandstones of the Frontier Formation along the Hogsback, Southwestern Wyoming. Des Plaines: Gas Research Institute, Tight Sands and Gas Processing Research Department, GRI-94/0020.

Lorenz, J.C., Sterling, J.L., Schechter, D.S., Whigham, C.L., and Jensen, J.L., 2002, Natural fractures in the Spraberry Formation, Midland basin, Texas: the effects of mechanical stratigraphy on fracture variability and reservoir behavior. American Association of Petroleum Geologists Bulletin, 86, 505–524.

Narr, W., 1996, Estimating average fracture spacing in subsurface rock. AAPG Bulletin, 80, 1565–1586.

Teufel, L.W., 1983, Determination of In-Situ Stress from Anelastic Strain Recovery Measurements of Oriented Core. Presented at SPE/DOE Low Permeability Gas Reservoirs Symposium, 14–16 March, Denver, Colorado.

Warpinski, N.R., Teufel, L.W., Lorenz, J.C., and Holcomb, D.J., 1993, Core Based Stress Measurements: A Guide to Their Application. Des Plaines: Gas Research Institute Topical Report GRI-93/0270, available through the Gas Technology Institute.

A2

Inclined Extension Fractures

Most extension fractures form as near-vertical planes because it takes less energy to open against the minimum compressive stress, which is typically horizontal, than against the other two stresses. The vertical weight of the overburden typically comprises the maximum compressive stress in structurally simple provinces, leaving the minimum and intermediate stresses in the horizontal plane. However, inclined stresses, and inclined extension fractures, are not uncommon.

When logging inclined fractures, one needs to define the reference for that inclination. In horizontally bedded strata, "inclined" fractures are inclined to both bedding and to vertical, but in folded strata fractures may be inclined to the vertical yet normal to the tilted bedding. These fractures are "high-angle" in a genetic sense, which can sometimes be used as evidence for fracturing prior to or during folding, but they are inclined fractures in the purely geometric sense.

Still other extension fractures are inclined relative to tilted bedding yet are vertical relative to the present-day up direction. These fractures probably formed in a conventional stress system where the weight of the overburden comprised the maximum compressive stress but after the strata had been folded.

If one is logging fractures strictly for inputs into a reservoir permeability model, it may be sufficient to log the present-day fracture dips, but if the purpose is to reconstruct the structural development of the strata then it is important to log fracture dip angles relative to both the vertical and the bedding.

A2a Inclined Extension Fractures in Horizontally Bedded Strata

Inclined extension fractures occur in some cores cut from unfolded, horizontal strata (Figures A2a.1–A2a.4), suggesting that compressive stresses are not always symmetrical about a vertical axis, even in basins that appear to be structurally simple.

Figure A2a.1 Narrow, inclined, calcite-mineralized extension fractures in a horizontally bedded shale. The fracture faces are marked with plume structures (not exposed in this view of the fracture edge). The apparent dip angles on the slab face are a few degrees shallower than actual dips due to the oblique intersection between the fracture planes and the slab face, but the fractures are not vertical. The vertical parting in the rock at the pencil point is a saw cut made for sampling, the horizontal breaks were created during core processing and handling. Slab face of a vertical, 4-inch diameter core; uphole is towards the top of the photograph.

Atlas of Natural and Induced Fractures in Core, First Edition. John C. Lorenz and Scott P. Cooper.

Figure A2a.2 A calcite-mineralized, inclined extension fracture, with faces marked by plume structure (not exposed in this view). The rock has parted along the fracture, showing that the true fracture dip is steeper than its apparent dip on the slab face. Vertical, 4-inch diameter core; uphole is towards the top of the photo.

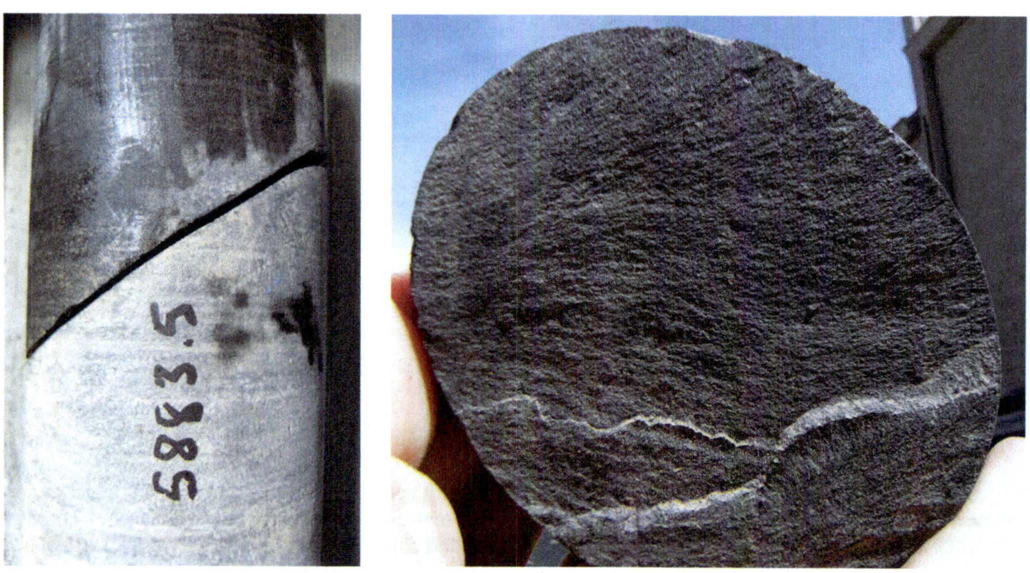

Figure A2a.3 Two views of an inclined extension fracture. The intermediate dip angle of the fracture plane (left) suggests shear, but the plume structure that decorates the fracture face (right) indicates an origin in extension. Vertical 3.5-inch diameter core; uphole is towards the top of the left photo, and away from the viewer in the right photo.

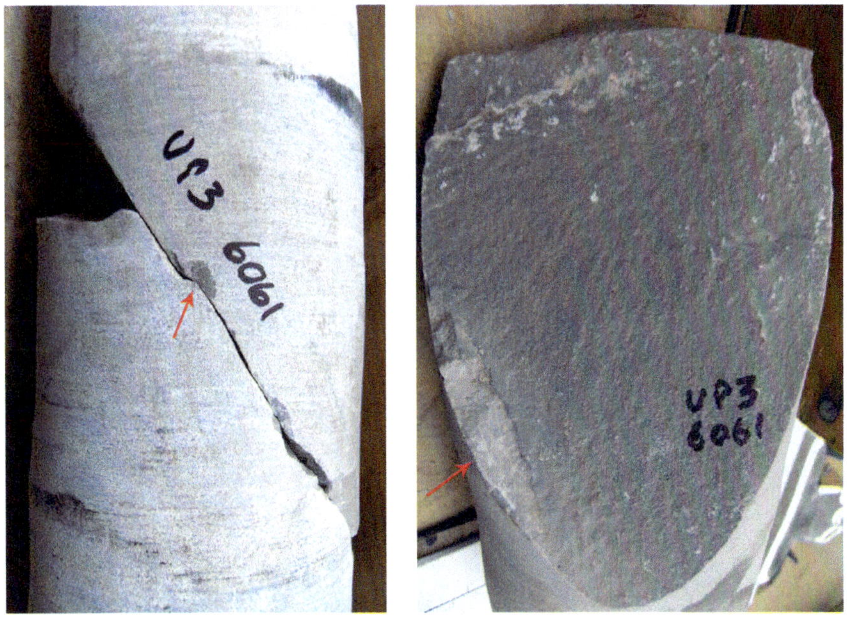

Figure A2a.4 Two views of a poorly mineralized, intermediate-angle extension fracture in a well-cemented sandstone. Left: this fracture plane might be mistaken for an inclined shear fracture, especially with the suggestion of a small step (*arrows*) in the middle of the fracture. However, the face of the fracture is marked by plume structure (right) indicating an origin in extension. The misleading step is where the rock broke other than along the fracture plane. The face of the fracture is largely unmineralized but it has a dark gray patina that contrasts with the lighter gray color of the rock exposed in a fresh break. This inclined extension fracture was formed by non-vertical stresses in this highly deformed reservoir. Vertical 4-inch diameter core; uphole is towards the top of the left photo, and upward and out of the plane of the right photo.

A2b Inclined Extension Fractures in Inclined Strata

Tilted, bed-normal fractures may have originally been vertical, becoming tilted with the host rock during folding (Figures A2b.1, A2b.2), or they may be the result of bed-normal extension during folding. In the latter case, they should strike parallel to the axis of folding. Extension fractures with horizontal dips are a special case and will be discussed separately.

Figure A2b.1 Bed-normal, calcite-mineralized, strata-bound extension fractures in silty units interbedded with calcareous shales. Inclined extension fractures can result when bed-normal, vertical fractures become tilted along with bedding during postfracture folding. Such fractures may be associated with indications of folding such as bed-parallel shear planes with slickenlines parallel to the dip azimuth of bedding. Vertical 4-inch diameter core; uphole is towards the top of the photo.

Figure A2b.2 Extension fractures in core cut from a limestone. The fractures are inclined relative to the vertical core axis. However, they are approximately normal to the prefold horizontal, which is defined by bedding and the stylolite, suggesting that they formed prior to tilting of the bedding. Vertical 4-inch diameter core; uphole is towards the top of the photo.

A2c Vertical Extension Fractures in Inclined Strata

Some high-angle extension fractures are vertical despite inclined bedding, suggesting that they formed after the beds had been tilted (Figure A2c.1).

Figure A2c.1 This narrow, calcite-mineralized extension fracture is parallel to the core axis and vertical, but it is oblique to bedding. This suggests that it formed after folding. Vertical 3-inch diameter core; uphole is towards the top of the photo.

A3

Horizontal Extension Fractures

Horizontal extension fractures are relatively rare in conventional reservoirs, but are common in some marine shale formations. Significant differences in mineralization and habit suggest that they are not the horizontal mechanical equivalent of vertical extension fractures. They are not easily distinguished from bedding on image logs and in fact, they were not generally recognized in the subsurface until long intervals of shale core began to be cut during the boom in shale and resource-play reservoirs.

A3a Beef-Filled Fractures

The most common type of horizontal extension fracturing occurs in marine shales, where prismatic calcite crystals are oriented normal to the fracture walls (e.g., Gale et al., 2014). The crystals are generally inferred to have grown outward from the walls, meeting at an irregular medial line marked by debris from the host rock (Figures A3a.1–A3a.5). This configuration is inferred to indicate that mineralization took place during, and kept pace with, the opening of the fracture.

Figure A3a.1 A calcite-filled horizontal extension fracture in a muddy marine shale. This fracture would be missed in an image log or on an unwashed outer core surface. Core commonly breaks along the horizontal fracture surfaces, as it has done here, and the entire fracture filling can be lost during core processing, leaving little if any evidence for fracturing. Vertical 4-inch diameter core; uphole is towards the top of the photo. The groove down the middle of the core surface is a core orientation line.

The origin of these structures is the subject of some debate. The conventional interpretation is that they are relatively shallow-burial, early-diagenetic features (e.g., Marshall, 1982; Al Aasm et al., 1992). In contrast, Cobbold and Rodrigues (2007) and Cobbold et al. (2013) suggest that these fractures formed when the strata were deeply buried under overpressured conditions.

Figure A3a.2 Two layers of mineralization consisting of prismatic calcite crystals that fill a horizontal fracture in a muddy shale. These long, narrow crystalline fibers are oriented normal to the fracture walls. The fibers have the same diameter along their length and are inferred to indicate that fiber growth was concurrent with and kept pace with fracture opening. The dark medial band consists of bits of the host rock and suggests that the fibers grew inward from the top and bottom fracture walls. Vertical 3-inch diameter core; uphole is towards the top of the photo.

Figure A3a.3 Two calcite-filled, horizontal extension fractures in a muddy marine shale: a thin bright white one in the middle of the photograph and a thicker, light gray one in the upper part of the photo on which the depth number is written. The calcite in the thicker fracture is white beneath the gray stain of the drilling mud, the gray stain suggesting that it has some permeability. Vertical 3-inch diameter core; uphole is towards the top of the photo.

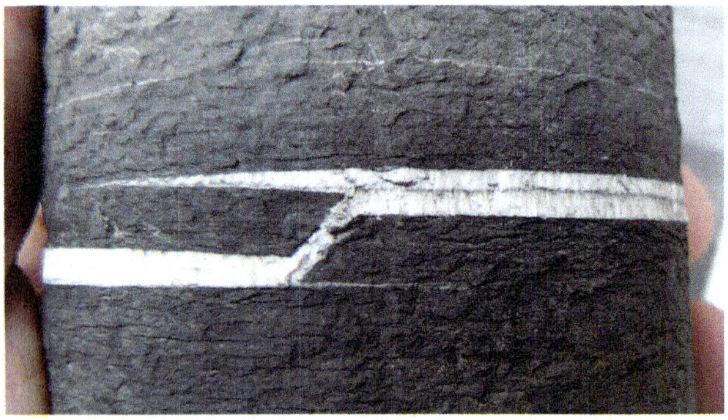

Figure A3a.4 Many horizontal, beef-filled fractures are not perfectly planar, but display diagonal offsets and wedging terminations. This fracture in a muddy shale has a medial line only where it is amalgamated at the right of the photo, suggesting that crystal growth in both of the leftward extensions was in one direction only rather than inward from both fracture faces. Vertical 3-inch diameter core; uphole is towards the top of the photo.

Figure A3a.5 Tilting of the white, calcite-mineralized extension fracture in concert with soft sediment deformation of the shale below a rapidly deposited calcarenite suggests that the bed-parallel fracturing in this photo formed before complete lithification. Vertical 3-inch diameter core; uphole is towards the top of the photo.

Figure A3a.6 The cross-cutting relationship (*arrow*) suggests that the high-angle, calcite-mineralized extension fracture at the left of the core cuts across and postdates the horizontal fracture at the bottom of the core, cut from calcareous shales. Vertical 3-inch diameter core; uphole is towards the top of the photo.

The horizontal, bed-parallel orientation of these fractures indicates that the minimum compressive stress at the time of fracturing was oriented vertically, and this is perhaps more easily explained mechanically if the fractures formed at shallow burial depths. The fracture faces are smooth and unmarked by plume structure, and fragments of the host rock material are locally included within the fracture mineralization, both compatible with incomplete lithification at the time of fracturing. Intersections between such horizontal fractures and high-angle extension fractures show that the horizontal fractures predate vertical fracturing (Figure A3a.6).

Although they are mineralized, such horizontal fractures may be more permeable than the nanodarcy-permeability shale reservoirs in which they are commonly found. Moreover, Rodrigues et al. (2009) have traced them for up to 100 m laterally in outcrops, thus they have the potential to connect the more steeply dipping types of fracture into an interconnected fracture network.

A3b Other Calcite-Mineralized Horizontal Extension Fractures

Some less well-defined calcite-filled, horizontal planes in cored shales (Figures A3b.1, A3b.2) may be beef-filled fractures that have been obscured by diagenesis and recrystallization, especially in older, more deeply buried formations.

Figure A3b.1 Calcite with translucent, equant crystals forms a millimeter-thick layer that cuts entirely across a vertical calcareous shale core (tan drilling mud obscures the right-hand side of the fracture surface). The origin of this layer is not obvious, but it occurs in a deeply buried Paleozoic formation and may be the recrystallized remnants of a fracture filled with prismatic calcite. Vertical 3-inch diameter core; view is towards the top of the core.

Figure A3b.2 Two views of a calcite-mineralized horizontal fracture. Top: the break in the middle of this silty core at the arrow is an obscure horizontal natural fracture of uncertain origin. Bottom: the calcite has a granular habit. This fracture would be missed on image logs, and would not necessarily be noted in lithologic logs of the core as the fracture is not obvious unless the core is picked up to expose the mineralized fracture surface. Much of the calcite filling broke away and was lost during the coring and handling processes. Vertical 3-inch diameter core; uphole is towards the top of the upper photo, and away from the viewer in the bottom photo.

A3c NOT Horizontal Extension Fractures

Several features, including organic shell material composed of prismatic calcite (particularly in Cretaceous marine strata), can form horizontal layers that resemble horizontal, beef-filled fractures (Figure A3c.1), particularly if they are interbedded with actual horizontal fractures. Shelly material is usually composed of less translucent prismatic calcite and may preserve organic linings or display growth ridges.

Bed-parallel shear planes can also resemble horizontal extension fractures unless distinguishing criteria such as slickenlines are observed. Near-vertical fractures can even resemble horizontal fractures if they are exposed only on a slab surface which was cut parallel to fracture strike, since the intersection of the fracture and slab planes can produce a horizontal calcite streak across the slab face. Horizontal induced disc fractures have also been mistaken for natural fractures in cores; disc fractures, described later in detail, are unmineralized and commonly marked with plume structures.

Some horizontal partings in core are filled with coal or oil. The former are usually coalified plant fragments (Figure A3c.2), whereas the latter are commonly created as oil bleeds out of the matrix onto bedding planes opened by the coring and handling processes. Some structures that appear to be horizontal extension fractures are sedimentary partings.

Figure A3c.1 Two views of a shell fragment composed of prismatic calcite in a silty Mesozoic marine shale. The shell material resembles a horizontal extension fracture filled with prismatic calcite. The absence of a medial line, the undulatory shell surface with remnant organic material, and the less translucent character of the prismatic calcite help distinguish this from beef-filled fractures. Vertical 3-inch diameter core; uphole is towards the top of both photos.

Figure A3c.2 A millimeter-thick horizontal layer of coaly material in a muddy marine shale. This is probably a coalified plant fragment since similar material with more definitive fern-like outlines is present in other parts of the same core. Such layers have been called hydrocarbon expulsion fractures, but there is no mechanical justification for that mechanism and no evidence to support its application to most of the fractures to which it is normally applied. Vertical 4-inch diameter core; uphole is towards the viewer.

References

Al-Aasm, I.S., Muir, I., and Morad, S., 1992, Diagenetic conditions of fibrous calcite vein formation in black shales: petrographic, chemical, and isotopic evidence. Bulletin of Canadian Petroleum Geology, 41, 46–56.

Cobbold, P.R., and Rodrigues, N., 2007, Seepage forces, important factors in the formation of horizontal hydraulic fractures and bedding-parallel fibrous veins ('beef' and 'cone-in-cone'). Geofluids, 7, 313–322.

Cobbold, P.R., Zanella, A., Rodrigues, R., and Løseth, H., 2014, Bedding-parallel fibrous veins (beef and cone-in-cone): worldwide occurrence and possible significance in terms of fluid overpressure, hydrocarbon generation and mineralization. Marine and Petroleum Geology, 43, 1–20.

Gale, J.F.W., Laubach, S.E., Olson, J.E., Eichhubl, P., and Fall, A., 2014, Natural fractures in shale: A review and new observations; American Association of Petroleum Geologists Bulletin, 98, 2165–2216.

Marshall, J.D. 1982, Isotopic composition of displacive fibrous calcite veins; reversal in pore-water composition trends during burial diagenesis. Journal of Sedimentary Petrology, 52, 615–630.

Rodrigues, N., Cobbold, P.R., Loseth, H., and Ruffet, G., 2009, Widespread bedding-parallel veins of fibrous calcite ('beef') in a mature source rock (Vaca Muerta Fm, Neuque´n Basin, Argentina): evidence for overpressure and horizontal compression. Journal of the Geological Society, London, 166, 695–709.

Section B

Shear Fractures

B1

Introduction

Shear fractures can accommodate both extension and shortening strains in host strata, typically in magnitudes greater than those accommodated by extension fracturing. More important to reservoirs, the effects of shear fractures on reservoir permeability are significantly different from the effects of extension fractures, since even small shear fractures with minimal, sub-millimeter-scale offset commonly form intersecting conjugate patterns with good interconnectivity whereas single sets of extension fractures form as parallel, non-intersecting planes. Shear fractures also have a greater potential for connecting a reservoir across lithologic boundaries since shear fractures are less affected by mechanical heterogeneity, and the face-parallel offset of shear fractures can break across the same thinner, relatively ductile beds that commonly arrest the growth of extension fractures.

Shear fractures with significant offset and slickensided surfaces are common and obvious in folded and faulted reservoirs, but shears with smaller offsets can also be found in less deformed strata where they are easily overlooked. Shear fractures may be more common than is generally recognized since millimeter-scale shear offset may not be obvious in a core and even large-scale shear can be difficult to recognize in image logs if the offset is parallel rather than normal to bedding. Shear can sometimes be inferred from the conjugate geometry of a shear fracture pair, but logs cannot image the slickenlined or stepped faces that definitively indicate shear where bedding offset is not apparent. Moreover, even if a conjugate fracture geometry is apparent in an image log, it may not be obvious whether the shear planes are open and permeable or are formed by the occluded, low-permeability, shear-related deformation bands.

Shear fracture widths and apertures tend to be less regular than those of extension fractures and the minor offsets of asperities often create voids that can significantly enhance reservoir permeability. However, it only takes a few millimeters of offset to damage the permeability of the host rock along a shear fracture face, filling in the voids with gouge and creating impermeable slickensided fracture faces. Additionally, shear and extension fractures form with different orientations relative to the in situ stress field and therefore the two fracture types do not have the same potential for shear, closure, and permeability alteration during production.

B1a Nomenclature

Shear fractures can have offsets of less than a millimeter, but offsets range upward to an undefined threshold where they are thought of as faults. In fact, some authors prefer to refer to any shear fracture as a "fault" but that broadens the definition of "fault," making it perhaps less valuable while leaving the useful term "shear fracture" without a home. On the other end of the scale, the geophysical community commonly uses the term "fracture" for any large planar structures that can be imaged seismically, using the term without genetic implications of shear or extension.

We find the term "shear fracture" to be useful in referring to discrete planes with shear offset, and prefer to use "fault" for larger shear structures with significant associated damage zones including secondary smaller shear fractures, fault breccia, and/or gouge. Fault zones also commonly include large-scale vuggy porosity within the associated breccia, as well as evidence for repeated shear and mineralization. We assign no threshold magnitude of offset to distinguish shear fractures from faults, and suggest rather that the terminology is similar to the vocabulary used for the continuous spectrum linking "puddle," "pond," and "lake." Such terms are useful in everyday parlance and easily understood despite the poorly defined and overlapping size implications, and the definitions can easily be refined for local use.

Atlas of Natural and Induced Fractures in Core, First Edition. John C. Lorenz and Scott P. Cooper.

B1b Anderson's Shear Fracture/ Fault Classification

Anderson (1905, 1942, 1951) theorized that, at least in homogeneous rock and in simple structural settings, steeply dipping shear fractures and faults should have strike-slip offsets while intermediate-angle shear fractures should have normal dip-slip offsets, and shallow, low-angle shears should have reverse dip-slip (thrust) offsets (Figure B1b.1). Moreover, he suggested that these geometries should form under specific stress conditions. Anderson's fracture/fault classification accommodates the mechanical implications of many cored shear fractures, and although not universally applicable we have found it useful. Briefly:

- Shear fractures with 60° dips (intermediate-angle dips) typically have normal dip-slip offset, recording a stress state where the maximum compressive stress was vertical and the minimum stress was horizontal, accommodating lateral extension normal to the strike of the fractures.
- Shear fractures with 90° dips (high-angle dips) are most often marked with evidence for strike-slip offset, recording a stress state where the maximum and minimum compressive stresses were in the horizontal plane, and accommodating lateral shortening oblique to fracture strikes.
- Shear fractures with 30° dips (low-angle dips) typically have indications of reverse dip-slip offset, indicating that the minimum compressive stress was vertical and the maximum compressive stress was horizontal. These shears accommodated lateral shortening by pushing the rock up small thrust-fault ramps and lifting the overburden.

As indicated by Figure B1b.2, the 30°, 60°, and 90° dip angles for the three types of shear are only the ideal; each type exhibits a range of dip angles when measured empirically in cores. Additionally, the ranges of the dip angles for the strike-slip, dip-slip, and reverse-slip shear categories overlap, reflecting lithologic heterogeneity and non-ideal stress conditions. In cores, high-angle shear fractures with indications of horizontal strike-slip offset such as slickensides, striae, and steps typically have dips steeper than 65°. Intermediate-angle shears with indications of normal offsets have dips of between 25° and 75°, and low-angle shears with reverse dip-slip indicators typically have dip angles of 40° or less. In addition, many shear fractures have surface indicators that record oblique slip and do not fit Anderson's categories.

Anderson's fracture classifications work best in minimally deformed strata, becoming less useful where the rocks have been significantly folded, tilted, twisted, or faulted since shear fractures can be tilted along with bedding so that the fractures no longer have their original dip angles or senses of shear. Early fractures can also be reactivated so that the latest-motion shear indicators are inconsistent with their geometries. Some shear planes show superimposed oblique slickenlines

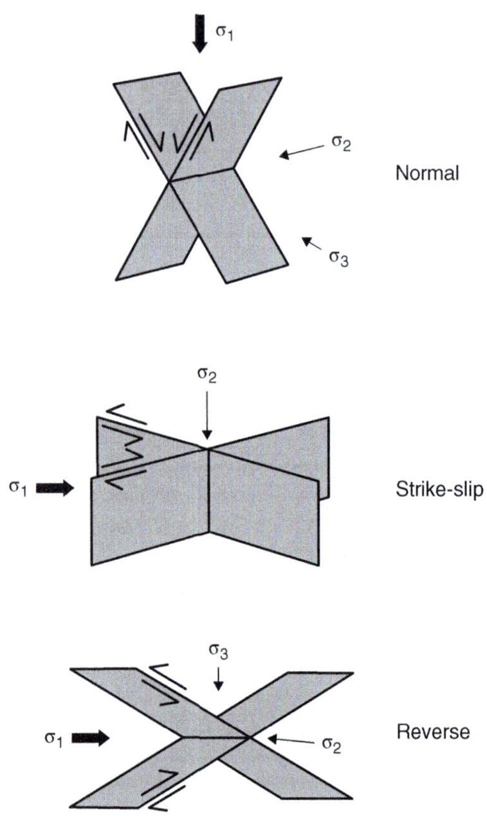

Figure B1b.1 Anderson's classification of shear fractures and faults, from Olsson et al. (2004), adapted from Anderson (1905, 1942).

SHEAR SENSE	DIP ANGLE		
	0°–40°	35°–75°	65°–90°
reverse dip-slip	A	R	R
normal dip-slip	R	A	R
strike-slip	R	C	A
oblique	R	C	C
bed-parallel	C, flexural slip, thrust ramps		
A	Andersonian, most common		
C	Common		
R	Rare		

Figure B1b.2 Observed ranges for the dip angles relative to vertical of strike-slip, dip-slip, and reverse dip-slip shear fractures in core in structurally simple settings.

recording multiple offset directions and events. The distinctions between Anderson's three categories become blurred or even inapplicable where the rock has a compound structural history. On the other hand, Anderson's categories can locally be used to determine age relationships between intersecting fracture systems (e.g., Olsson et al., 2004).

Shear fractures commonly form as conjugate pairs. The ideal intersection angle for a conjugate pair is 60° but conjugate pairs with larger or smaller angles are common (e.g., Hancock and Bevan, 1987). The conjugate nature of fractures in cores can easily be missed unless the core is oriented or unless long intervals of the core are carefully pieced together. However, cores provide only a small sampling of a formation and shear fractures may be present in cores that did not sample the complimentary set of the conjugate pair. Moreover, many conjugate fracture pairs are less than systematic, with fractures of one half of the pair dominating at any given locale (e.g., Seyum and Pollard, 2012; Shainan, 1950), and the absence of an intersecting conjugate set in a core or image log does not mean that an intersecting pair is not present in the larger reservoir.

Nevertheless, a conjugate geometry can be part of the identification criteria for a shear fracture system and in fact, it may be the only evidence for shear available from an image log. Care must be taken in making this interpretation, since similar intersecting geometries can be formed by superimposed extension fractures sets. Additional criteria such as bedding offsets, shear-related fractography, and the orientations of the fractures relative to the in situ stresses should be used where possible to support an interpretation of shear.

While the Anderson model of pairs of intersecting conjugate shear planes works well where extension was in one direction only (plane-strain shear), some shear-fracture systems consist of an orthorhombic arrangement of two conjugate shear pairs that suggest extension in two directions (i.e., Krantz, 1989; Reches and Dieterich, 1983; Sagy et al., 2003) (Figure B1b.3). Such systems, or oblique shear-plane reactivation in structurally complicated settings, may account for multiple shear pairs in some systems and for raking slip directions in others, but

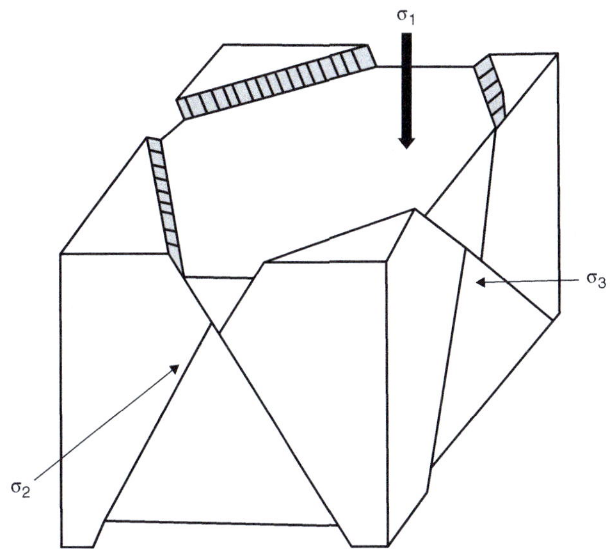

Figure B1b.3 If shear accommodated extension in two directions rather than just one, as in some rift basin settings (e.g., Sagy et al., 2003), two shear pairs with an orthorhombic configuration are likely to develop. The three axes noted are the principal (1), intermediate (2), and least (3) compressive stresses (σ) (from Reches and Dieterich, 1983).

documenting such systems from the minimal samples provided by core is difficult.

Bed-parallel shear also does not fit into Anderson's classification. It commonly forms as flexural slip in folded, thin-bedded strata, with strain being accommodated along bedding planes, the classic analogy being shear between the pages of a folded telephone book. Slickenlines on the flexural slip fracture surfaces of such systems are usually parallel to the dip azimuth of the folded bedding, but they may have been rotated or reactivated in oblique shear by subsequent deformation. Bed-parallel shears may also form as links between low-angle thrust shears, created by small-scale versions of the ramp and flat geometry common in thrust belts.

Bed-parallel shears follow mechanically weak bedding contacts, and the dip angles are determined by the tilting of bedding rather than by Anderson's mechanical criteria; dip angles do not indicate a mode of origin for these fractures.

References

Anderson, E.M., 1905, The Dynamics of Faulting. London: Geological Society, Special Publication, vol. 367, pp. 231–246.

Anderson, E.M., 1942. The Dynamics of Faulting and Dyke Formation with Application to Britain. Edinburgh: Oliver and Boyd.

Anderson, E.M., 1951, The Dynamics of Faulting and Dyke Formation with Applications to Britain, 2nd edn. Edinburgh: Oliver and Boyd.

Hancock, P.L., and Bevan, T.G., 1987, Brittle modes of foreland extension, *in* Coward, M.P., Dewey, J.F. and Hancock, P.L. (eds) Continental Extensional

Tectonics..London: Geological Society, Special Publication, vol. 28, pp. 127–137.

Krantz, R.W., 1989, Orthorhombic fault patterns: the odd axis model and slip vector orientations. Tectonics, 8(3), 483–495.

Olsson, W.A., Lorenz, J.C., and Cooper, S.P., 2004, A mechanical model for multiply-oriented conjugate deformation bands. Journal of Structural Geology, 26, 325–338.

Reches, Z., and Dieterich, J.H., 1983, Faulting of rocks in three-dimensional strain fields, 1. Failure of rocks in polyaxial, servo-control experiments. Tectonophysics, 95, 111–132.

Sagy, A., Reches, Z., and Agnon, A., 2003, Hierarchic three-dimensional structure and slip partitioning in the western Dead Sea pull-apart. Tectonics, 22(1), 1004.

Seyum, S., and Pollard, D.D., 2012, Echelon crack geometries in limestone, and associated shear localization. Stanford Rock Fracture Project, 23, A-1–17.

Shainan, V.E., 1950, Conjugate sets of en echelon tension fractures in the Athens Limestone at Riverton, Virginia. Geological Society of America Bulletin, 61, 509–517.

B2

Shear Fracture Dimensions

Shear fracture heights are less well constrained by the data available from core than are the heights of extension fractures since they are commonly inclined relative to the axis of a core, which therefore captures only a small sample of each fracture. On the other hand, and for the same reason, a numerically larger sampling of the fractures of a shear set is likely to be captured by a core.

Shear fractures are more likely than extension fractures to cut across bedding boundaries and through multiple lithologic/geomechanical units because the magnitude of rock displacement associated with shear can be much greater than that of extension fractures. As a result, shear fracture heights are poorly constrained by core data even while they are more likely to enhance vertical connectivity within stratified reservoirs. As with extension fractures, shear fracture lengths are essentially unconstrained by core data.

Unmineralized shear fracture surfaces are typically rougher than unmineralized extension fractures due to the steps and en echelon segments that form along shears, and due to offsets of these fracture-face irregularities against each other. Shear fracture apertures are discontinuous, so that fluid-flow pathways along shear planes are tortuous channels around the offset asperities and through the adjacent hollows. Because the original shear surface is rough and irregular, mineralization in shear fractures tends to be discontinuous. Hollows along a shear fracture may be lined with crystalline mineralization indicating mineral growth into open void spaces, while the adjacent high points on the same fracture are unmineralized due to contact with the opposing fracture face. More importantly, a shear fracture that has a high flow capacity within and along the fracture aperture may also have faces lined with slickensides that inhibit flow from the matrix into the fracture aperture (see Nelson, 2002).

The common occurrence of shear fractures as conjugate pairs means that shear fracture strikes often have more variability than those of extension fractures (Figure B2.1). Fracture strike measurements must also be supplemented by measurements of fracture dips in order to fully assess the geometry of dip-slip and reverse dip-slip shear-fracture networks; the parallel strikes portrayed for such fractures on rose plots resemble strikes for extension fractures, but the opposing dip azimuths and intersecting dip angles become apparent on stereonets.

Fracture spacing is a common input parameter for fluid-flow models and is easily measured in horizontal cores that contain the parallel planes of an extension fracture set. Measurable spacings are less useful where a fracture system consists of two sets of intersecting conjugate shear fractures. Spacings normal to fracture faces for each set of the conjugate pair typically have a log-normal distribution (Figure B2.2), but a measured

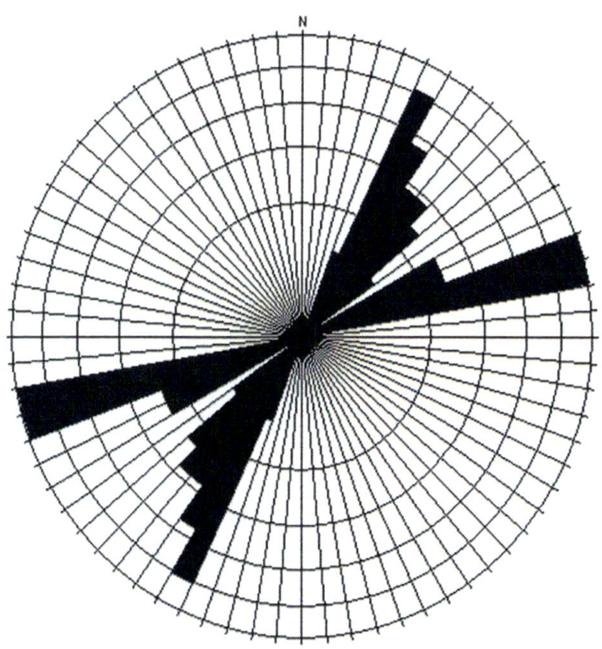

Figure B2.1 The strikes of a pair of strike-slip conjugate shear fractures, occurring together in a suite of horizontal cores cut from Paleozoic marine sandstones of the Spraberry Formation, West Texas (n = 56) (from Lorenz et al., 2002).

Atlas of Natural and Induced Fractures in Core, First Edition. John C. Lorenz and Scott P. Cooper.

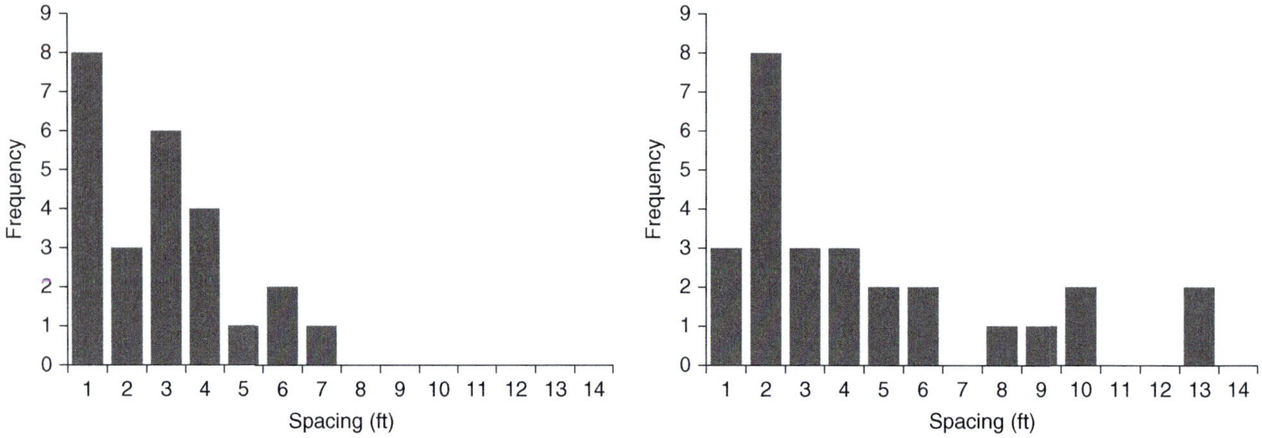

Figure B2.2 Spacings (feet), normal to the fracture planes, of NNE-SSW striking right-lateral and ENE-WSW striking left-lateral strike-slip conjugate shear fractures (left and right histograms, respectively), occurring together as a conjugate, strike-slip shear pair in horizontal cores cut from Paleozoic marine sandstones of the Spraberry Formation, West Texas (data from tables presented in Lorenz et al., 2002).

spacing between intersecting fractures is useless since it changes depending on how close to the point of intersection the core cuts the two fracture planes. Nevertheless, an estimate of reservoir block sizes enclosed by intersecting fractures can be obtained from quantitative core data.

References

Lorenz, J.C., Sterling, J.L., Schechter, D.S., Whigham, C.L., and Jensen, J.L., 2002, Natural fractures in the Spraberry Formation, Midland basin, Texas: the effects of mechanical stratigraphy on fracture variability and reservoir behavior. AAPG Bulletin, 86, 505–524.

Nelson, R.A., 2002, Geologic Analysis of Naturally Fractured Reservoirs, 2nd edn. Boston: Gulf Professional Publishing.

Olsson, W.A., Lorenz, J.C., and Cooper, S.P., 2004, A mechanical model for multiply-oriented conjugate deformation bands. Journal of Structural Geology, 26, 325–338.

B3

Shear Fracture Fractography

Shear fracture faces can be distinguished by a variety of characteristics including slickensides, slickenlines, en echelon segmentation, tool marks, glassy surfaces, bedding offsets, and both accretionary and non-accretionary steps. Petit and Laville (1987) made a detailed study of shear fracture ornamentations and their interpretations (see also Doblas, 1998), but only a subset of that variety is common in the cores cut from hydrocarbon reservoirs since most reservoirs are in less deformed structural settings. Mineralization can also record shear, appearing as sheared crystals ("slickencrysts"; see Hancock, 1985) when offset is large, or as a stepped, asymmetric, fish-scale texture when offset is small. Mineralization may also only appear to have been sheared if it has been precipitated onto a sheared fracture face, preserving an impression of the sheared and slickenlined fractography. Some shear fracture apertures are obstructed by gouge or small-scale breccia created during shear.

B3a Slickensides, Slickenlines, and Accretionary Steps

The term "slickensides" refers to glassy, metamorphic, striated fracture surfaces created by significant shear offset where a high normal stress pressed the fracture faces together during shear (Figures B3a.1, B3a.2). A slickensided surface also commonly displays low-amplitude steps created by the accretion of gouge material during shear (Figure B3a.3). The asymmetric steps are oriented such that the steep step risers would seem to act as one-way ratchets to prevent motion if shear were to be reversed, so the direction of last offset on a shear surface can be inferred from the striations and the sense of that motion can be determined from step asymmetry.

The similar term "slickenlines" usually refers to a linear pattern of striations on a fracture face without surface metamorphism (Figure B3a.4). Slickenlines form under conditions of smaller offset and/or a lower magnitude of stress normal to the fracture face. They also form during shear in poorly lithified rock. Slickenlines show the direction of last motion on the face but with a 180° ambiguity if there are no steps to constrain the sense of motion.

Figure B3a.1 Slickensides on a horizontal shear plane in a well-cemented quartz sandstone show that the direction of last offset was parallel to the double-headed arrow. The associated asymmetric accretionary steps show that the missing block moved to the lower right. Vertical, 3-inch core; uphole is away from the viewer.

Figure B3a.2 Two views of slickenlined, gouge-filled fractures in a sandstone. The photo on the right is of the surface indicated by the arrow in the photo on the left. The white, non-calcareous gouge is a clayey material consisting of sand grains that were ground into rock flour during shear.

Figure B3a.3 Slickenlines record oblique strike-slip offset on a high-angle shear fracture in sandstone. Subtle asymmetric accretionary steps suggest that the sense of motion was left-lateral. Patches of gouge are preserved on the fracture face. Vertical 4-inch diameter core; uphole is towards the top of the photo.

Figure B3a.4 Slickenlines record dip-parallel offset on an intermediate-angle inclined shear plane in a gray shale. The fracture surface is mineralized with a muddy, rust-red calcite. The sense of shear (normal or reverse) is indeterminate from the fractography but normal offset is mechanically more likely than reverse dip-slip offset given the 50° dip angle of the fracture plane. Vertical 4-inch diameter core; uphole is towards the viewer.

B3b En Echelon Segments

Some shear fractures accommodated small-scale offsets through the formation of en echelon fracture segments within a zone of shear (Figures B3b.1, B3b.2) rather than along a more continuous, discrete shear plane, usually because the rock was relatively ductile at the time of fracturing. En echelon fracture segments are usually considered to be short extension fractures within a zone of shear. The en echelon fracture segments have also been interpreted as riedel shears oriented at low angles to the overall trend of the shear zone, although they rarely show evidence of shear. Conjugate en echelon fractures are most common in limestones, but they can occur in any lithology that fractured under temperature and pressure conditions that made the rock relatively ductile.

Some fractures only appear to be segmented, the segments joining into a single plane in the third

Figure B3b.1 Left-stepping, calcite-mineralized segments of a strike-slip shear fracture in a muddy sandstone. The shear zone containing these segments strikes approximately 40° counter-clockwise to the maximum horizontal compressive stress as recorded by the strike of associated induced petal fracture ("PF," parallel to the red line) and comprise one set of a conjugate pair that is present in the core. Vertical, 4-inch diameter core; uphole is towards the viewer.

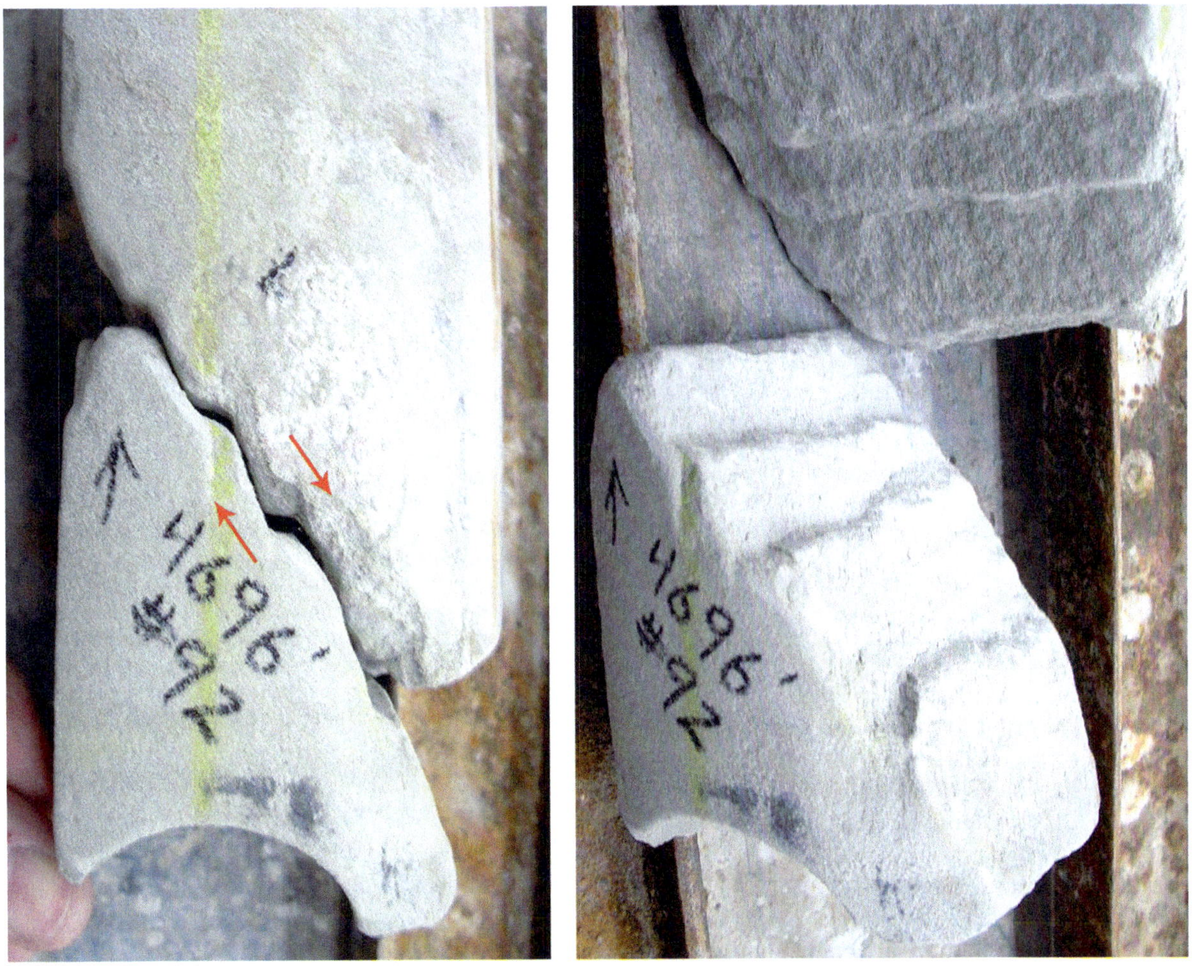

Figure B3b.2 Two views of en echelon segments of an intermediate-angle, dip-slip shear fracture system in a clean sandstone. The zone enclosing the stepped segments comprises one leg of a conjugate X formed by two shear zones, with parallel strikes and opposing dips. The upper core piece moved downward with minimal, normal, dip-slip offset relative to the lower piece of core. The short, abrupt risers between the steps were not formed by the shear process but by breakage of the core between the en echeon segments. The bisector of the inferred conjugate angle is vertical. Vertical 4-inch diameter core; uphole is towards the top of the photo.

Figure B3b.3 En echelon, overlapping fracture segments that indicate a narrow shear zone associated with intermediate-angle, dip-slip shear fracturing in an oil-soaked, poorly cemented sandstone. The bracket indicates the width of the shear zone in which the fracture segments formed, and the inferred bisector of the conjugate angle is vertical. Vertical 4-inch core; uphole is towards the top of the photo.

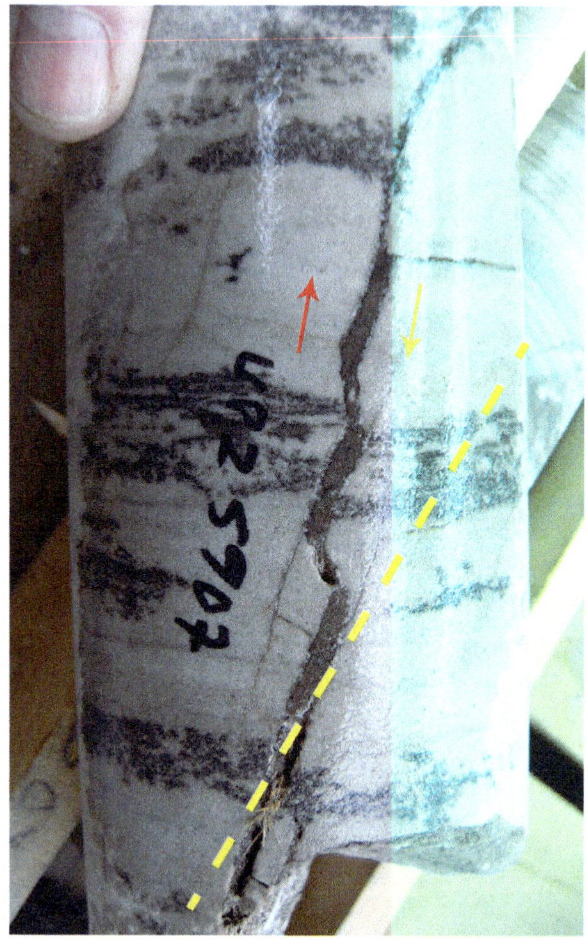

Figure B3b.4 En echelon, overlapping fractures forming a fracture system within a shear zone in a sandstone. Continued shear broke the rock between the segments, opening wider slots and connecting them. Irregular oil stains mark offsets of the bedding, but they are only approximate indicators of the magnitude of offset. The fracture slots are partially mineralized with oil-stained calcite. The fibrous material that is stuck in the open aperture near the bottom of the photo is Lost Circulation Material (commonly referred to by the initials LCM and probably consisting of cedar bark) that was pumped into the hole during coring in order to reduce the loss of drilling mud into the fracture system. The sense of offset of the en echelon segments is incompatible with a conjugate pattern that would form one leg of an upright X with a vertical conjugate bisector, but this example comes from a highly deformed structural domain and the X pattern may be tilted such that the bisector of the X lies along the trend indicated by the yellow dashed line, rather than vertical and parallel to the core axis. Vertical 4-inch diameter core; uphole is towards the top of the photo.

dimension. Such patterns are typical of twist hackle on an extension fracture, so caution must be used in making shear interpretations based on en echelon segments. The step-overs of en echelon shear segments will also step consistently in one direction for one fracture set of the conjugate pair, and in the other direction for the other set.

Segments of the set that strikes clockwise from the bisector of the acute conjugate angle should step to the right, and segments of the other set should step to the left. If the segments step in the other directions, then the segmented fractures may be inconsistent with an en echelon shear pattern (Figure B3b.4). Some en echelon segments may be oriented nearly parallel to the general trend of a narrow shear zone, and the segments in other systems are highly oblique to a wider zone of shear, depending on stress ratios and the ductility of the rock at the time of fracturing (e.g., the convergent and divergent en echelon vein arrays of Smith, 1996).

B3c Steps

Shear fractures where the millimeter-scale offsets were insufficient to form slickensides, slickenlines, or en echelon segments commonly have faces that are covered with small, low-amplitude steps (Figures B3c.1–B3c.3). The long treads

Figure B3c.1 Two views of conjugate, intermediate-angle, dip-slip shear fractures in an arkosic, coarse-grained, fluvial channel sandstone of the Pennsylvanian Abo Formation, central New Mexico. Left: viewed parallel to strike the fractures from dip-slip conjugate pairs. Right: the faces of the left-dipping fractures display stepped surfaces that record small amounts of normal dip-slip shear where the bisector of the conjugate angle and the maximum compressive stress were vertical. The missing block in the right photo moved upward by no more than a millimeter, *against* the asymmetry of the steps on the fracture face.

Figure B3c.2 Two views of an intermediate-angle, normal-offset, stepped, shear fracture in a sandstone core. The steps record small-scale shear offset with the core segment to the right of the fracture ("NF") moving down relative to the core segment left of the fracture. The sense of offset of these steps and the vertical bisector to an inferred conjugate pair are the same as for the en echelon segments illustrated by Figure B3b.2, in the same core, indicating that they are part of a spectrum of features of different size but similar origin. The pinch and swell structures described below illustrate the other, small-scale end of this spectrum of structures. Vertical 4-inch core; uphole is towards the top of both photos.

Figure B3c.3 A stepped, high-angle, strike-slip shear fracture face in horizontal sandstone core. The sense of shear indicated by the steps is left-lateral, with the missing block moving away from the viewer. Horizontal $2^{5/8}$-inch core; stratigraphic up is towards the top of the photo, uphole is away from the viewer.

Figure B3c.4 Continued shear on a stepped shear-fracture surface can lead ultimately to slickensides and gouge, but if there is only a little continued shear, especially if the normal stress across the fracture face is low, the high points of the steps can be slicked without affecting the lows between steps. High-angle, strike-slip shear fracture in oil-soaked, cross-bedded sandstone. Vertical 4-inch diameter core; uphole is towards the top of the photo.

of the steps are essentially small-scale, unconnected, en echelon fracture segments, and the short risers are the recently broken connectors between the steps, similar to but smaller than those just illustrated and with the same sense of asymmetry. The asymmetry of these steps relative to the direction of offset is opposite to that of the steps found on slickensided fractures and formed of accreted, comminuted rock created by larger shear offsets.

If shear continues without a high normal stress holding the fracture closed, the fracture-face asperities can become polished or even slickensided (Figure B3c.4). Continued shear beyond that produces a more pervasive slickensided surface.

B3d Pinch and Swell

"Pinch and swell" describes a subtle pattern commonly related to minimal offset along a stepped shear fracture when viewed in a cross-section normal to the fracture plane. Where offsets are of the order of a centimeter, the irregularities of the fracture plane can easily be mentally reassembled to preshear positions (Figure B3d.1), but the pattern is less obvious where the scale of offset is oblique to the plane of view, or where offset is of the order of a few millimeters (Figures B3d.2–B3d.4) or of a scale that can only be detected with a hand lens (Figure B3d.5).

Small shears that exhibit pinch and swell textures are relatively common, but they are also commonly tightly cemented and the stepped surfaces are therefore rarely exposed. Where exposed, the fracture faces (Figure B3d.6) display asymmetric steps that are smaller but otherwise similar to the larger shear fracture steps just described.

Pinch and swell can occasionally be found along parallel fracture planes (Figure B3d.7). Since shear fractures typically form as conjugates rather than as parallel planes, this suggests shear reactivation of a set of earlier, parallel extension fractures.

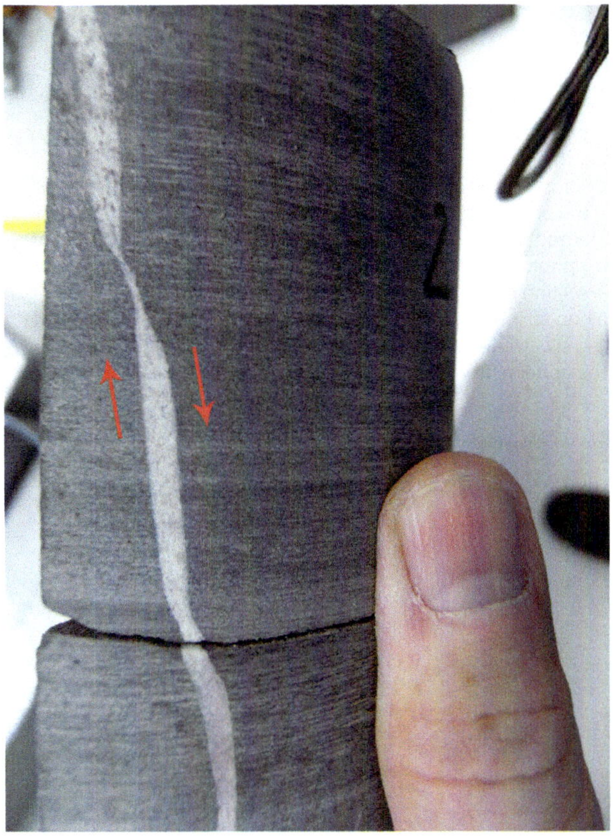

Figure B3d.1 A large-scale pinch and swell pattern viewed in section across the fracture plane. The fracture is filled with quartz and the host rock is a quartz-cemented quartz sandstone. High-angle shear fracture with dip-slip offset of the order of 2 cm. Butts from a 2.5-inch vertical core; uphole is towards the top of the photo.

Figure B3d.2 Small-scale, calcite-mineralized, pinch and swell textures along intersecting conjugate shear planes in a calcite-cemented sandstone. These high-angle fractures have oblique-slip offsets and are from a structurally complex setting on the thrusted forelimb of an anticline. Vertical 4-inch diameter core; uphole is towards the top of the photo.

Figure B3d.3 Two expressions of compatible small-scale, high-angle, strike-slip shear in a limestone. The two multi-stranded fractures (center and left) are en echelon segments accommodating right-lateral shear within narrow shear zones; the pinch and swell texture at the right records the same sense of shear but it is more localized and it accommodated slightly more offset. Somewhere in the larger system beyond the core, one would expect to find a complementary left-lateral fracture set, striking about 60° clockwise relative to these fractures. Vertical 4-inch core; uphole is towards the viewer.

Figure B3d.4 Intermediate-angle dip-slip shear is recorded by the calcite-mineralized pinch and swell pattern exposed on the surface of this core cut from a calcareous shale. The complementary shear set, with parallel strike and opposing dip, is not present in this core. Vertical 4-inch diameter core; uphole is toward the top of the photo.

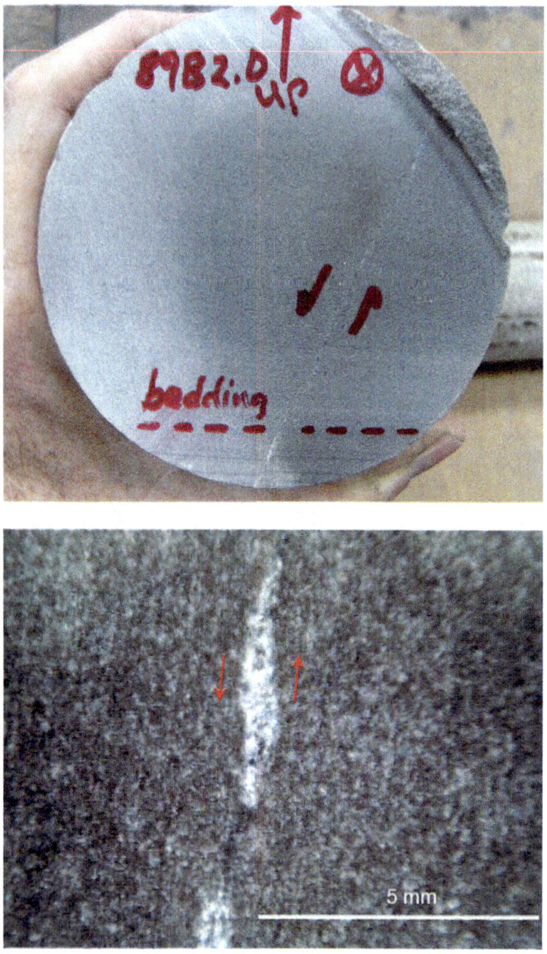

Figure B3d.5 Two examples of obscure, calcite-cemented, small-scale pinch and swell textures along intermediate-angle, normal dip-slip shear planes. Top: shear in a fine-grained calcite-cemented sandstone. The magnitude of offset can be seen in the displaced bedding couplet at the bottom of the photo. See Figure B3d.6 for the surface fractography of this fracture. Horizontal 4-inch diameter core; stratigraphic up is at the top of the photo, uphole is away from the viewer. Bottom: photomicrograph showing a section of the repetitive pattern of small-scale pinch and swell textures along an intermediate-angle, dip-slip shear fracture in a massive, fine-grained siltstone. Vertical 4-inch diameter core; uphole is oblique, towards the upper left of the photo.

Figure B3d.6 A rare view of the stepped surface of a pinch and swell shear fracture with millimeter-scale offset. The fracture is cemented with calcite (*white*) but the asperities that touch the opposing fracture face are unmineralized (*gray*). This is one of the small-scale shear fractures in the same core shown in the top photo in Figure B3d.5. Horizontal 4-inch diameter core; stratigraphic up is at the top of the photo, uphole away from the viewer.

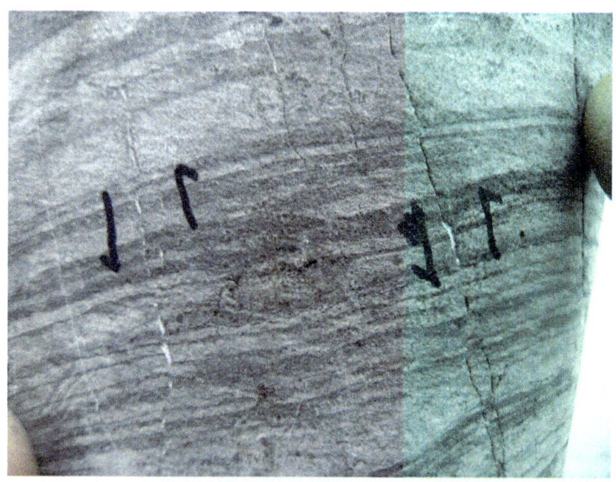

Figure B3d.7 Numerous parallel shear planes with consistent small-scale offset down on the left in a tilted, laminated shaley limestone. Parallel shear planes do not often form as primary structures, so these may be parallel extension fractures that were reactivated and connected in shear. One plausible mechanical interpretation is that bed-parallel shear during folding created secondary shear within the beds, tilting and reactivating an early-formed suite of extension fractures. Vertical 4-inch diameter core; uphole is towards the top of the photo.

B3e Sheared and Glassy Surfaces

Clays on the faces of a shear plane in muddy rock can be aligned by shear, leaving a seemingly polished, mirror-like surface that is often striated (Figure B3e.1). Minor offsets of bedding and fossils indicate that these surfaces can develop with surprisingly little displacement, less than a few millimeters. Some of these fractures are mineralized with veneers or pockets of calcite (Figures B3e.1, B3e.2) indicating that there has been permeability and fluid flow between the polished fracture faces, although the fracture width may now be plugged or at least reduced by that mineralization. Other shears are unmineralized (Figure B3e.3). Water droplets left on the glassy surfaces evaporate before they imbibe into the rock, indicating the low permeability of the fracture faces.

Figure B3e.1 A mineralized, polished, and striated low-angle fracture surface in a muddy shale. The striations (parallel to the dashed red line) are oblique to the dip and strike of the shear plane, and it does not fit easily into one of Anderson's classifications, suggesting structural complications during or after the fracture plane formed. A veneer of translucent calcite mineralization, white where it is peeling away from the fracture face (red arrow), filled the original fracture aperture. Vertical 4-inch diameter core; uphole is towards the top of the photo.

Figure B3e.2 A near-horizontal polished and striated shear plane in a muddy shale. Lineations indicate left-right offset. The low-relief surface has a slight asymmetry with striated, left-facing inclines that suggest the missing block moved to the right. Crystalline calcite has been precipitated to the right of right-facing inclines, filling hollows left by a few millimeters of offset. Vertical 4-inch diameter core; uphole is towards the viewer.

Figure B3e.3 A striated, near-horizontal shear plane in a muddy shale. The layer of rock that was striated and metamorphosed by shear is about half a millimeter thick and is flaking off the underlying host rock, revealing the undeformed, structureless mudstone below. Vertical 4-inch diameter core; uphole is towards the viewer.

B3f Slickencrysts

The mineralization within shear fractures and faults commonly also shows an asymmetrically stepped habit with lineations called "slickencrysts" (mineral fibers or crystals aligned by shear and forming acute angles with the fracture wall; Hancock, 1985). Bedding offsets show that the asymmetry of the steps indicates the sense of shear, i.e., the steep, abrupt faces of the steps face in the direction of movement of the missing block (Figure B3f.1).

Slickencrysts suggest that the mineralization was sheared during or after precipitation. If the mineralization is calcite and the calcite is twinned, it was probably strained and sheared after it was deposited in the fracture. Sheared calcite tends to be white, opaque, and

Figure B3f.1 Two views of slickencrysts along an intermediate-angle, dip-slip shear fracture in a muddy shale. The missing block in both photos moved downward with a sense of normal shear as indicated by both bedding offset and the asymmetry of the steps developed in the calcite. Vertical 2-inch diameter core; uphole is towards the top in both photos.

amorphous whereas unstrained calcite deposited in the fracture after the shear occurred is more commonly clear, translucent, and crystalline.

B3g Other Evidence for Shear

Where only the edge of a fracture is exposed or where the fracture face fractography is obscure due to dissolution, mineralization, drilling mud, etc., there may be other evidence to indicate shear. Evidence such as offset bedding (Figure B3g.1), breccias along the shear plane, and trapezoidal pull-apart voids formed between pairs of shear planes (Figures B3g.2, B3g.3) can be obvious. Less obvious evidence includes seemingly non-systematic yet characteristic patterns of irregular fractures that form when a set of systematic, high-angle extension fractures is reactivated and connected by secondary shear (Figure B3g.4).

There is, of course, no guarantee that the plane on which a shear fracture is exposed offers the best view of the offset, and in fact, it is often difficult to detect and determine offset if the direction of shear was in and out of the plane of view. As always, the core must be picked up and examined on all sides in order to get complete, three-dimensional views of cored fractures.

Figure B3g.1 Shear in a core cut from chalk is recorded by offsets on both the bedding couplet near the top of the photo and the brown inclined fracture cutting across the middle of the photo (below the arrow). The apparent senses of the two offset are mechanically impossible if offsets are vertical and parallel to the slab. However, the curved shape of the shear plane highlighted by the arrow suggests that much of the offset was in and out of the plane of the photograph. Moreover, although not apparent in this 2D exposure, the brown fracture plane cutting upper right to lower left across the photo is inclined both down to the left and down and towards the viewer, so strike-slip shear along the central wedge of chalk results in apparent vertical offset where none exists. Vertical core, quarter-core remnant of a 4-inch diameter core; uphole is towards the top of the photo.

Figure B3g.3 A small-scale pull-apart along a minor high-angle strike-slip shear fracture. Vertical 4-inch diameter core; uphole is towards the viewer.

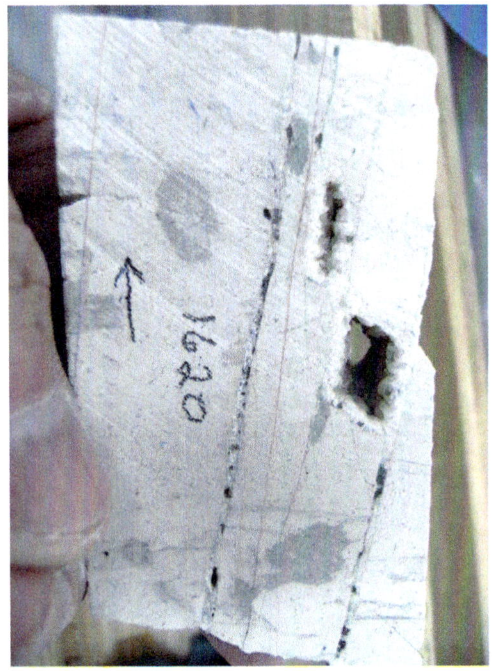

Figure B3g.2 Two trapezoidal pull-apart voids lined with calcite crystals, in a chalk. These structures formed by vertical shear offset along the near-vertical fractures combined with parting and extension along subhorizontal planes that are probably related to bedding. Similar features can form with shear along bedding planes and the opening of trapezoidal voids at bed-normal extension fractures. Vertical core, quarter-core remnant of a 4-inch diameter core; uphole is towards the top of the photo.

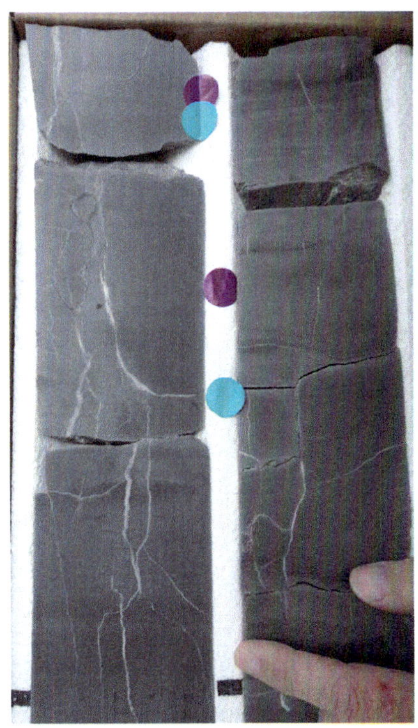

Figure B3g.4 An irregular fracture system in fine-grained limestone, composed of (1) relatively planar, high-angle extension fractures (fracture strike at an oblique angle to the slab plane exaggerates the fracture irregularity), and (2) curved and low-angle shear fracture connections between the extension fractures Shear was superimposed on the extension fracture system (see Figures C8.3, C8.4 for another example). Slabs of 4-inch diameter vertical core; uphole is towards the top of the photo.

References

Doblas, M., 1998, Slickenside kinematic indicators. Tectonophysics, 295, 187–197.

Hancock, P.L., 1985, Brittle microtectonics: principles and practice. Journal of Structural Geology, 7, 347–457.

Petit, J.P., and Laville, E., 1987, Morphology and microstructures of hydroplastic slickensides in sandstones, *in* Jones, M.E., and Preston, R.M.F., eds, Deformation of Sediments and Sedimentary Rocks. Palo Alto: Blackwell Scientific, pp. 107–121.

Smith, J.V., 1996, Geometry and kinematics of convergent conjugate vein array systems. Journal of Structural Geology, 18, 1291–1300.

B4

High-Angle Shear Fractures

B4a Introduction

In Anderson's classification, high-angle shear fractures are strike-slip shears that should form when the maximum compressive stress is in the horizontal plane and the vertical stress created by the weight of the overburden is only the intermediate compressive stress. In an ideal, homogeneous mechanical medium, high-angle shear fractures would all have vertical dip angles and show horizontal, strike-slip offset. However, for the following examples we consider a fracture to be a high-angle strike-slip shear if it has a dip angle of greater than 65° and the slickenlines on the fracture face are within 15° of horizontal. There are many examples in structurally complex settings where offset on a near-vertical shear fracture is oblique rather than horizontal, where horizontal offset occurs on fractures with dip angles of less than 65°, or where a strike-slip fracture that formed with a near-vertical dip has been tilted along with bedding and no longer has a high dip angle relative to vertical even though it is still normal to bedding.

As with high-angle extension fractures, high-angle shears have a low probability of being intersected by vertical cores since the fracture planes are parallel to the core axis (see Figure A1c3.3). The inclined planes of normal and reverse dip-slip shear fractures have much higher chances of being captured by vertical cores and image logs, the probability of intersection depending on fracture dip angles and intensities. Intersection probabilities for inclined fractures and deviated wellbores also depend on wellbore azimuth and deviation, as well as fracture strike and dip, and calculations for determining the relative abundance of different fracture sets from one-dimensional wellbore data become poorly constrained as wellbores and fracture planes deviate from vertical or horizontal orientations.

Determinations of right-lateral vs. left-lateral senses of offset in cored fractures use the same criteria used in outcrop (i.e., steps, chatter, offsets, etc.; see Petit and Laville, 1987; Ramsay and Huber, 1983).

B4b High-Angle Strike-Slip Shear Fractures

Ideal high-angle shear planes have near-vertical dips and display indications of horizontal offset (Figures B4b.1, B4b.2). If horizontal slickenlines or lineations have been obscured by mineralization or removed by dissolution, a variable dip angle combined with a uniform strike can be evidence for strike-slip offset (Figures B4b.3, B4b.4) since extension fractures are typically planar rather than curved and since dip variations do not impede offset in the horizontal plane. Breccia may also indicate shear where the fracture surface cannot be inspected (Figure B4b.5).

Atlas of Natural and Induced Fractures in Core, First Edition. John C. Lorenz and Scott P. Cooper.

Figure B4b.1 A slickenlined, high-angle strike-slip shear plane in a muddy siltstone cuts the core in half vertically, leaving the core in numerous pieces due to additional core-handling breaks along bedding. The core is from a continental-scale strike-slip tectonic setting. Drilling mud has been washed off the fracture surface in the core piece in the foreground but obscures much of the rest of the fracture. Vertical 4-inch diameter core; uphole is away from the viewer.

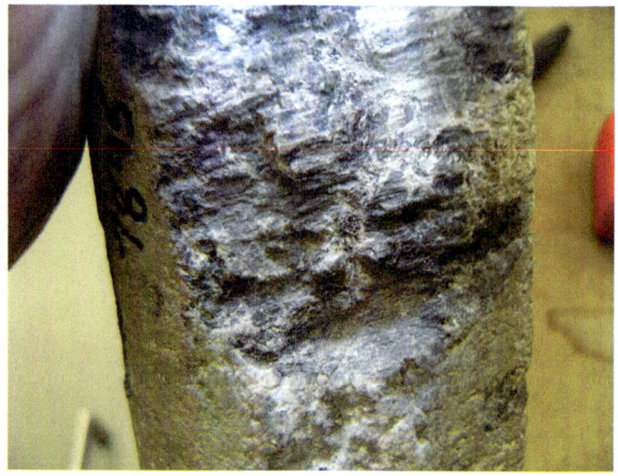

Figure B4b.2 The accretionary asymmetric steps on a high-angle shear plane in dolomite indicate right-lateral strike-slip offset. The core is from a well located in front of a thrust belt where strike-slip and reverse dip-slip shears would be expected due to the high horizontal, thrust-normal compressive stress. Vertical 4-inch diameter core; uphole is towards the top of the photograph.

Figure B4b.3 Two views, profile (left) and face (right), of an irregular high-angle strike-slip shear plane in the butts of core from an oil-soaked eolian sandstone. Strike-slip shears may be irregular in the vertical cross-section as shown on the left, but they are linear in plan-view cross-sections where they had to accommodate horizontal offset. The slightly inclined slickenlines and accretionary steps suggest left-lateral offset. Vertical 4-inch diameter core; uphole is towards the top of both photographs.

Figure B4b.4 Two views of a high-angle shear fracture in the slabs of core cut from a calcite-cemented eolian sandstone. Left: the fracture plane (arrow) is irregular with a variable dip angle but, within the confines of the core, it is planar in horizontal cross-section, suggesting that it is a strike-slip shear even though no surface fractography is preserved. This interpretation is supported by the irregular fracture width despite a host lithology that is not susceptible to dissolution, and the presence of other, more obvious shear fractures elsewhere in the core. (Individual fractures should not be interpreted as stand-alone structures but rather fracture sets in a core should be interpreted using as many mutually supporting characteristics as possible.) The oil-stained calcite of the mineralized fracture face opposite the arrow is shown in the right-hand photo; muted, poorly defined crystals suggest minor dissolution of the calcite mineralization. Vertical 4-inch diameter core; uphole is towards the top of both photographs.

Figure B4b.5 An irregular high-angle fracture exposed on the slab face of a core cut from a fine-grained dolomite. Breccia indicates shear but the sense of offset is not obvious in this plane. The steep dip angle and an absence of offset bedding suggest horizontal offset in and out of the plane of the photo. From a broadly folded and faulted anticline where repeated core runs came up short indicating heavily fractured and faulted strata. Vertical 4-inch diameter core; uphole is towards the top of the photograph.

B4c Non-Ideal High-Angle Shear Fractures

Core is rarely held in its in situ position when being examined in the laboratory, and a left-lateral shear exposed on a horizontal plane cut across a vertical core (Figure B4c.1) will appear to be a right-lateral shear when the core is held upside down, as it often is when that position offers a better view of the fracture. Moreover, some apertures that appear to be related to pull-aparts along an irregular shear fracture surface may in fact be dissolution slots.

High-angle shear fractures in structurally complex settings may not conform to the Andersonian model. Figure B4c.2 shows shear fractures with relatively steep but variable dip angles and opposing dip azimuths, suggestive of a dip-slip conjugate shear pair, but the slickenlines on the fracture are highly oblique, raking about 45° across the fracture surface. Other fractures have false slickenlines formed by sedimentary layering rather than shear offset (Figure B4c.3).

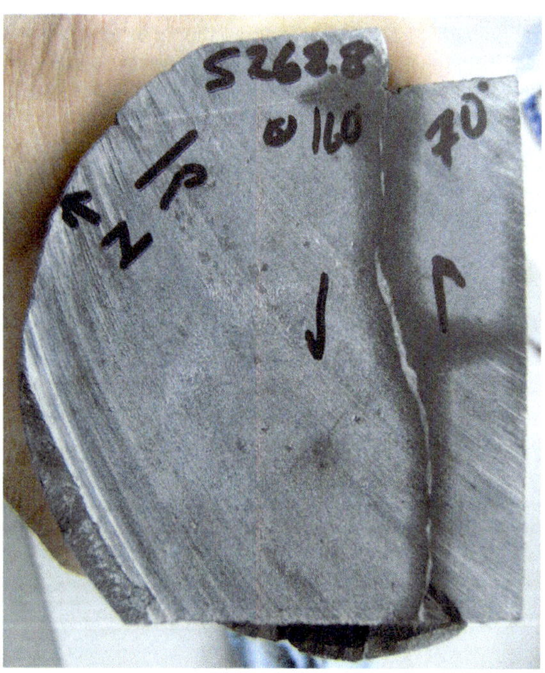

Figure B4c.1 Apparent left-lateral offset on a high-angle shear fracture exposed on a saw cut across core from a fine-grained limestone. The aperture is irregular due to dissolution along the fracture plane and possibly initiated at a void opened by oblique offset. The circled dot just left of "160°" indicates that this is a map view of the core, looking downhole. If this was a view of the bottom of the core segment looking uphole, the sense of offset indicated by the shear arrows would remain the same but the interpretation would be right-lateral. This is oriented core, and the fracture strikes marked on the core (160°, 70°) were calculated using the orientation survey. Vertical 4-inch diameter core; uphole is towards the viewer.

Figure B4c.2 Two views of a high-angle shear fracture with oblique shear recorded by inclined slickenlines (upper right to lower left in the right photo) on the fracture face. The left photo shows a curved fracture surface forming the left side of the core. The right photo shows the fracture face, where calcite crystals were deposited in the hollows formed between offset asperities, and the asperities were dragged against the opposing faces to form slickenlines. An unidentified black material, possibly the remnants of an early oil charge in the reservoir, covers much of the fracture face. Vertical 4-inch diameter core; uphole is towards the top of both photos.

Figure B4c.3 Pseudo-lineations on an irregular, high-angle extension fracture mimic strike-slip shear but are caused by the coarse-fine laminations of cross-beds of an eolian sandstone. Vertical 4-inch diameter core; uphole is towards the top of the photograph.

References

Petit, J.P., and Laville, E., 1987, Morphology and microstructures of hydroplastic slickensides in sandstones, *in* Jones, M.E., and Preston, R.M.F., eds, Deformation of Sediments and Sedimentary Rocks. Palo Alto: Blackwell Scientific, pp. 107–121.

Ramsay, J.G., and Huber, M.I., 1983, The Techniques of Modern Structural Geology, 1: Strain Analysis. London: Academic Press.

B5

Intermediate-Angle Shear Fractures

Ideal, Andersonian intermediate-angle shear fractures and faults have 60° dip angles as well as slickenlines that record normal dip-slip offset. The actual geometries of intermediate-angle shear fractures in cores, where dip-parallel slickenlines or stepped faces (Figures B5.1, B5.2) occur on shear planes having dip angles ranging from about 35° to 75°, are not as tightly constrained. Intermediate-angle shears commonly occur in conjugate pairs with parallel strikes and opposing dip azimuths (Figure B5.3), although this may not be apparent unless lengths of core are removed from the core boxes and locked together end to end. The direction of offset recorded by slickenlines on the shear surfaces of intermediate-angle cored shear fractures is typically within ±15° of parallel to the dip azimuth.

Ideal intermediate-angle shears form where the weight of the overburden comprises the maximum compressive stress. As with high-angle shears, however, these fractures can be found in structurally complex settings that are unrelated to the conditions of initial fracturing, and the trends of slickenlines on shear planes with intermediate dip angles can record oblique-slip and even strike-slip offset (Figures B5.4, B5.5). An inclined fracture plane with little or no fractographic markings is most plausibly interpreted as a shear given an intermediate dip angle and a simple structural setting (Figure B5.6), but as shown in Chapter A2, intermediate-angle fractures can also form in extension. Some shear fractures record a compound history of deformation, and fracturing is not readily interpreted from the superimposed, multi-faceted evidence (Figure B5.7).

Intermediate- and low-angle shears cut oblique to the axes of vertical wellbores, so they have a high probability of being captured by a core or an image log. The probability of intersecting these inclined fractures with a deviated or horizontal wellbore is also high but is also governed by fracture strike relative to the wellbore azimuth.

Intermediate-angle compaction shears that form prior to lithification are common in muddy strata and can resemble the more planar, intermediate-angle shears formed under higher stress conditions after lithification. However, compaction shears, described later in Chapter C7, typically have undulatory surfaces and non-systematic strikes.

Atlas of Natural and Induced Fractures in Core, First Edition. John C. Lorenz and Scott P. Cooper.

Figure B5.1 Two views of an intermediate-angle, unmineralized shear plane in a muddy shale. The shear plane has a 48° dip angle and is marked with dip-parallel, dip-slip slickenlines. Clay particles become aligned parallel to the fracture face during shear to form a polished surface. Vertical, 4-inch diameter core; uphole is towards the top of both photos.

Figure B5.2 Two views of an intermediate-angle shear plane in an indurated, calcite-cemented sandstone. The left photo shows an edge-on profile of the slightly undulatory fracture plane with a centimeter-scale pinch and swell texture, constraining the magnitude of offset. The arc outlined with a dashed line below the shear fracture is an induced petal fracture. The right photo shows the stepped surface of this fracture, with slightly oblique dip-slip slickenlines on the dark asperities, interspersed with white patches of calcite precipitated in the hollows between asperities. Vertical, 4-inch diameter core; uphole is towards the top of the left photo, and away from the viewer in the right photo.

Figure B5.3 Two views of intermediate-angle shear planes in a massive mudstone. Two shears, marked by blue arrows in the left-hand photo, have parallel strikes as indicated by their near-horizontal expressions on the slab plane at the left side of the core in the left photo, and opposing dip azimuths. The fracture surfaces are marked by faint dip-slip slickenlines and a thin veneer of calcite, obscured by drilling mud (right-hand photo). Vertical, 4-inch diameter core; uphole is towards the top of both photos.

Figure B5.4 Two views of an intermediate-angle shear fracture, showing an edge-on view of the inclined plane and small secondary shear fractures in the intact rock (left), and calcite slickencrysts (right) that record an unexpected strike-slip offset. The missing block moved to the viewer's right. Vertical, 4-inch diameter core; uphole is towards the top of both photos.

Figure B5.5 An intermediate-angle shear fracture in a muddy sandstone, with a measured dip angle of 50° and slickenlines showing last motion raking across the fracture face at a 60° angle to the dip azimuth. Vertical, 4-inch diameter core. View is downhole, with uphole obliquely out of the plane of the photo and towards the viewer.

Figure B5.6 Some intermediate-angle inclined fractures have indeterminate origins. This inclined fracture in a muddy limestone offers neither bedding offsets nor fractographic evidence that would help determine its origin. Many times such fractures can be plausibly interpreted based on more definitive evidence found in similar and/or parallel fractures in the same core. This fracture occurs in a moderately deformed structural setting at the crest of an anticline and is probably a dip-slip shear fracture with minimal offset. Vertical, 4-inch diameter core; uphole is towards the top of the photo.

Figure B5.7 Irregular shear fracturing in a slightly metamorphosed sand-shale sequence. The main shear plane is opened in a combination of shear and extension, offsetting bedding by a few millimeters. The overall trend of the fracture is oblique to bedding as well as inclined relative to vertical. 2-inch diameter core. Bedding is tilted and the hole is inclined. Uphole and stratigraphic up are unknown.

B6

Low-Angle Shear Fractures

Idealized Andersonian low-angle shear planes have 30° dip angles with reverse (thrust) offset parallel to the dip azimuth. As with kilometer-scale thrust faults, core-size shears record a horizontal compressive stress that exceeded the vertical stress at the time of failure, lifting the overburden along low-angle ramps. Since low-angle shears cut across the axis of a vertical wellbore, they have a high probability of being captured by a core or image log, and they seldom occur in isolation in a core.

Although not abundant, a surprising number of low-angle shears occur in cores cut from relatively simple structural settings. Moreover, many cored low-angle shears accommodated only a few millimeters or centimeters of offset, suggesting that shear failure and minor offset raised the stress level in the system and arrested further fracture propagation.

Low-angle shear fractures can be hard to recognize in cores and image logs since they resemble bedding and since sedimentary rocks offer few vertical features that can be used to document offset (Figure B6.1). Core ends and fracture faces must be examined for the slickenlines and steps that indicate shear (Figures B6.2, B6.3), and sometimes it is necessary to break a core along a suspected low-angle shear plane in order to gain enough evidence to make the correct interpretation. Reactivation of low-angle shears, recorded by superimposed oblique slickenlines on a fracture face, is common in some structural domains.

Low-angle shears can occur as conjugate pairs in a core (Figures B6.4, B6.5), but the core must be examined carefully and it must be locked together in continuous intervals in order to recognize the diagnostic parallel strikes and opposing dips. Low-angle shearing may follow optimally oriented inclined bedding and sometimes follows the fore-sets of well-developed cross-beds. These structures can provide one of the conjugate sets of a pair, with the complementary conjugate set cutting irregularly oblique to bedding (Figure B6.6). Some low-angle planar features in cores are not marked by definitive fractography (Figure B6.7), but are interpreted as shear planes based on similarities to other, less ambiguous shears in the same core or to conjugate geometries with similar planes.

Figure B6.1 A well-developed pinch and swell pattern and an offset sedimentary structure in the lower part of this photo document reverse dip-slip shear along a small-scale, calcite-mineralized, low-angle shear fracture in a muddy carbonate. A nearly parallel but less regular shear plane cuts across the core in the upper third of the photo. Slab from a vertical 4-inch diameter core; uphole is towards the top of the photo.

Atlas of Natural and Induced Fractures in Core, First Edition. John C. Lorenz and Scott P. Cooper.

Figure B6.2 Two views of a low-angle shear plane in a bioturbated silty shale. Slickenlines show that shear was parallel to the dip azimuth, but the sense of shear (normal or reverse) cannot be determined from the fractography. Reverse slip is mechanically the most plausible sense of offset on a low-angle shear plane unless the shear occurred in undercompacted sediments or was part of a large-scale gravitational detachment surface. This shear plane in the core butts is poorly expressed in the equivalent slabs of the core which were badly broken up by the slabbing process. The oblique exposure of the shear fracture on the slab plane produced an apparent dip that is nearly parallel to the irregular bedding; the presence of this shear plane would be easy to miss if only the slab face were to be inspected. This sample is from a structural domain that is presently in regional extension, suggesting either local structural complications or previous horizontal compression. Vertical, 4-inch diameter core; uphole is towards the top of both photos. The groove down the slab face is from a mechanical "scratch" test that measures present-day mechanical properties of the rock.

Figure B6.3 Stepped slickencrysts on a low-angle shear plane in a limestone. The missing block moved to the left. This sample is from a mildly folded zone several kilometers in front of a fold and thrust belt. Vertical, 4-inch diameter core; uphole is away from the viewer.

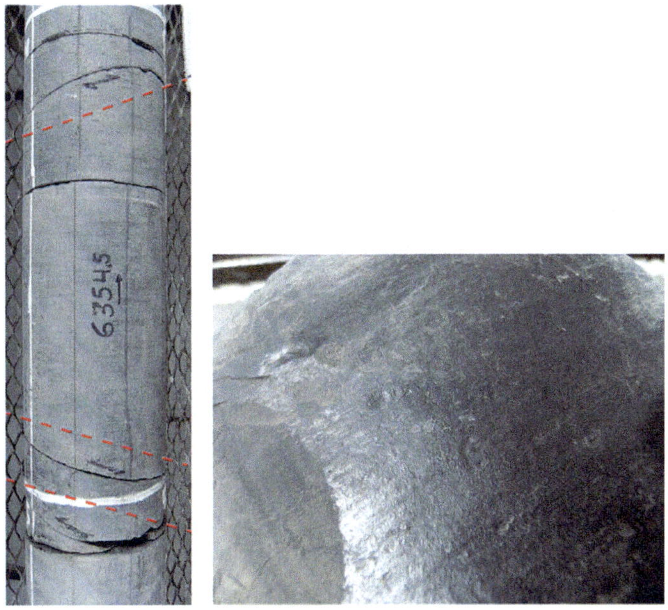

Figure B6.4 Left: opposing conjugate shear planes in a siliceous shale, parallel to the dashed red lines and as marked by the blue double-headed arrows drawn on the core surface. Bedding is horizontal, normal to the core axis. The white line across the core is a depth marking. The right photo shows the lightly slickenlined surface of one of these shear fractures. This sample is from a structurally simple domain. Unslabbed, vertical 4-inch diameter core; uphole is towards the top of the left photo, and oblique, out of the plane of the photo and upward in the right photo.

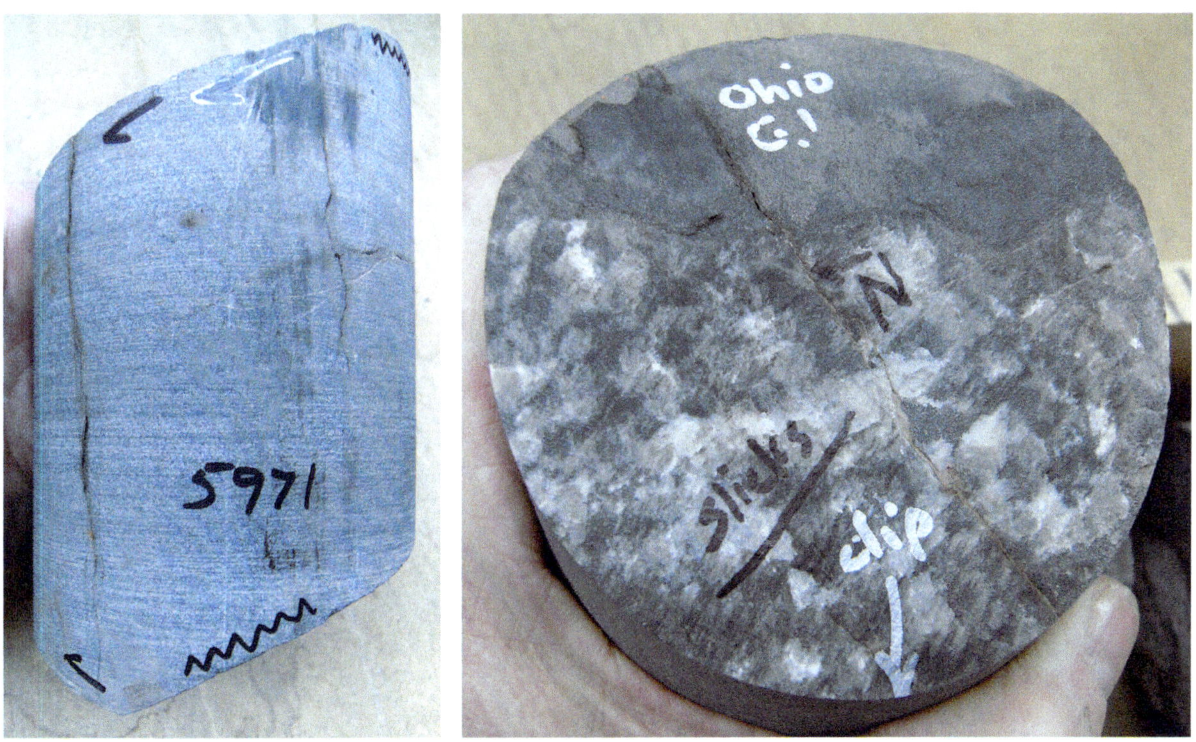

Figure B6.5 Two views of two low-angle shear planes with oblique slickenlines and slickencrysts in core cut from a well-cemented sandstone. Left: the two shear planes marked by half arrows drawn on the core have parallel strikes and opposing dip azimuths. The zig-zag line at the base of the core is one of a series of cross-bed foreset bedding planes, one of which was sheared to form the upper shear of the conjugate pair. Right: the slickenlines on one of the series of shear surfaces in this core are parallel to the black line ("slicks") marked on the core and are oblique to the dip azimuth. This example is from a tightly folded anticline with local faults. Vertical, 4-inch diameter core; uphole is towards the top of the left photo, and away from the viewer and obliquely downward in the right photo.

Figure B6.6 Two views of a conjugate shear fracture pair in a core cut from a muddy limestone. One of the conjugate shear planes (B) follows a clay parting in the tilted bedding; it is marked by slickenlines as shown in the right photo. The opposing conjugate shear plane (A) cuts oblique to bedding and is less regular, being marked by stepped slickencrysts. Neither shear plane accommodated significant offset. This core is from a fold and thrust setting where reverse-shear conjugates would be expected. Slab (left) and equivalent butt (right) of vertical, 4-inch diameter core; up is towards the top of the left photo, and away from the viewer in the right photo.

Figure B6.7 A low-angle inclined plane (left) with a nondescript surface (right) in a siliceous shale. Other than bedding, the most plausible origin of a low-angle break in a core is shear, but in this example there is little to prove it except possible faint slickenlines parallel to the trend of the dip azimuth. Vertical 4-inch diameter core; uphole is towards the top of the left photo and away from the viewer in the right photo.

B7

Bed-Parallel Shear Fractures

Bed-parallel shear fractures, where fracture geometry is controlled more by the orientation of the bedding planes than by the magnitudes and orientations of the three compressive stresses, do not provide much information on the in situ stress conditions. Bed-parallel shears that formed as flexural slip planes are common in folded, bedded shales, as well as in folded, thinly bedded formations that are composed of alternating competent and incompetent strata. Shear within the incompetent units and at their contacts with competent units requires less energy than other forms of strain when accommodating folding in such strata. Where the cored strata are tilted, these shears are both bed-parallel and low-angle relative to vertical (Figures B7.1, B7.2). Thus, it is important to relate shear planes to local bedding attitudes as well as to vertical in order to make proper mechanical interpretations. Bed-parallel shears can also form as connecting structures between low-angle inclined thrust ramps in thrust terrains (Figures B7.3, B7.4), similar to the larger-scale ramp and flat geometry common in thrust belts,

Figure B7.1 Two views of bed-parallel shear planes. Left: three bed-parallel shear planes (parallel to the dashed red lines) follow shale partings in a thin-bedded, folded limestone cored on the flank of an anticline. Right: the polished and slickenlined surface of one of the shear planes, with the recorded sense of motion parallel to the trend of the dip azimuth as would be expected of flexural-slip shear. The shear surface also shows hints of a secondary, oblique set of slickenlines. Butts of vertical 4-inch core; uphole is towards the top of the left photo, and towards the viewer in the right photo.

Atlas of Natural and Induced Fractures in Core, First Edition. John C. Lorenz and Scott P. Cooper.

Figure B7.2 Two views of a slickenlined, unmineralized, bed-parallel shear plane in a variably calcareous shale with inclined bedding. Left: the shear plane, just below the double-headed arrow, as exposed on the slab face is obscure and easily missed. Right: the slickenlines on the fracture face show that the bed-parallel break is a shear plane. Slabs of vertical, 3-inch diameter core; uphole is towards the top of the left photo and towards the viewer in the right photo.

Figure B7.3 Two views of a subtle, polished and slickenlined, lightly calcite-mineralized shear plane in a muddy shale with horizontal bedding. Left: the break in the core just below the double-headed arrow drawn on the core is a shear plane. Right: slickenlines and calcite mineralization deposited in hollows document shear along this plane but would be missed if the ends of the core were not examined. The core was cut from lightly deformed strata in front of a fold and thrust belt. Vertical, 4-inch diameter core; up-hole is towards the top of the left photo, and obliquely upward and out of the plane of the right photo.

Figure B7.4 Two views of a bed-parallel shear plane at a bentonite layer in a marine shale. Shear is not obvious from the outer surface of the core. Much of the bentonite layer and all evidence for shear were lost when this core was slabbed. The core is from a structurally simple area, but such slickenlines can form with only a few centimeters or even a few millimeters of offset. Vertical 4-inch diameter core; uphole is towards the top of the top photo, and upward and obliquely towards the viewer in the bottom photo.

Figure B7.5 Bed-parallel shears can be quite subtle, as this one from a gently tilted, bedded anhydrite on an anticlinal fold. The anhydrite unit is pervasively and plastically deformed in some zones but contains numerous discrete bed-parallel shears in others, all recording fold-related bed-parallel flexural slip that is more obvious higher in the core where it cut a limestone-shale formation. Minimal shear offset is recorded by stepped shear surfaces (right photo). Slabs of vertical, 4-inch diameter core; uphole is towards the top of both photos.

although the origin of such fractures is unclear if the transition between ramp and flat has not been cored. The origin of bed-parallel shears in flat-lying strata is sometimes obscure (Figures B7.5, B7.6). Minor bed-parallel shears can also form as complements to bed-normal stylolitization, where the removal of material along the stylolite requires slip along the bedding planes separating that bed from the adjacent (above and below) un-shortened strata. The reverse can also happen, where opening of an extension fracture in one bed requires bed-parallel slip along contacts with the adjacent strata.

Figure B7.6 Two views of a quartz-mineralized bed-parallel fracture in an ancient, well-cemented, siliceous siltstone. Low-grade metamorphism and recrystallization of the rock have largely destroyed evidence for shear on this fracture face, but other, similar fractures in this core show subtle slickenlines. Vertical 4-inch diameter core; uphole is towards the top of both photos.

Bed-parallel shear is hard to detect in image logs, where it resembles bedding unless it is accompanied by significant brecciation or mineralization. Bed-parallel shear can also be easy to miss in core since most core studies look only at the slab surfaces where bed-parallel shears with little or no mineralization look like run-of-the-mill induced bed-parallel breaks.

Bed-parallel shears provide clues for reconstructing the structural development of a reservoir. They also offer potential connectors for fluid flow between high- and intermediate-dip-angle fractures, creating an interconnected fracture network that would go unrecognized if only the more obvious high-angle fractures in a core are logged.

B8

Deformation Bands

Deformation bands are shear planes that form in high-porosity, poorly cemented rock. They are common in eolian sandstones (Figures B8.1, B8.2) (Aydin and Johnson, 1978; Jamison and Stearns, 1982), but have also been recognized in high-porosity sandstones deposited in other environments (Figure B8.3) and in chalks (Rath et al., 2011). Nelson (2002) suggests deformation bands are most likely to form in a sandstone where porosity exceeds 18%. These structures can be mineralized where they have been resheared after the rock became more competent, but typically they are not mineralized, consisting rather of planes of collapsed porosity and crushed grains that create locally reduced permeability (Figure B8.4).

Deformation bands are fractures in the mechanical sense as they form planes in the rock that were created by shear. However, deformation bands do not enhance permeability. Rather, they form baffles and barriers to fluid flow in what would otherwise be good-quality reservoir rock (Antonellini and Aydin, 1994; Fossen et al., 2007). Nelson (2002) recognized these structures in his early work on eolian sandstones, referring to them as "gouge-filled fractures" although "deformation band" has become the more commonly accepted term. The rock for a centimeter or so adjacent to a deformation band is commonly more compacted and more competent than the host rock.

Deformation bands are not planes of mechanical weakness, thus deformation band surfaces are rarely exposed. Rare exposures of deformation band surfaces in outcrop display subtle steps similar in form and asymmetry to those described in section B3c, and continued shear along these zones can create slickenlines.

Deformation bands may consist of isolated shear planes or they may form irregular, multi-stranded bundles (Figure B8.5). Some bands are planar and effectively unlimited in length, others consist of multiple en echelon segments that are each centimeters to decimeters in length. Shear offset along each band is typically of the scale of millimeters to a few centimeters, but systems composed of many tens of amalgamated deformation bands can accommodate greater offset. Some deformation band systems form Andersonian conjugate geometries (Olsson et al., 2004), others form as mingled, irregular strands and ladders. They can form as tails off fault asperities (see Figure B8.3), or as highly irregular, seemingly non-systematic networks (Figure B8.6).

Deformation bands are easily mistaken for mineralized fractures in cores, and they are difficult to distinguish from other fracture types in image logs. The true nature of deformation bands can be especially difficult to assess where oil migration has saturated the core, as capillary forces tend to concentrate the oil along the small, comminuted grains in the shear zone, obscuring the structure (Figures B8.6, B8.7). Thin sections provide definitive characterizations of deformation bands, showing the diagnostic reduced porosity and crushed grains. Identification of deformation bands in core is important since the effect of a system of deformation bands in a reservoir is very different from that of most other fracture types.

Atlas of Natural and Induced Fractures in Core, First Edition. John C. Lorenz and Scott P. Cooper.

Figure B8.1 Left: a deformation band in an oil-stained eolian sandstone. The sense of offset (a few millimeters down on the left side of the photo as seen in bedding offsets near the top of the photo) is obscured by the concentric, near-horizontal bands circling the core surface created by drill-bit rotation, and by irregular oil staining of the inclined cross-bed foresets (dipping down to the right). Measurements from the aggregate set of deformation bands in this core show that the bands form a set of a parallel-striking, opposite-dipping, conjugate pairs. Right: a close-up photo of the top of a different multi-stranded deformation band in the same core, showing the band as it exits the side of the core. What looks like mineralization is actually comminuted sand grains. Vertical, 4-inch diameter core; uphole is towards the top of both the photos.

Figure B8.2 An inclined deformation band (arrow marked "DB") in a high-porosity eolian sandstone. The deformation band offsets bedding by about a centimeter, down on the right side of the band. The irregular, more vertical shear fracture with open remnant void space ("F") has only a few millimeters of offset. Offsets of the fracture and the deformation band are combined where they intersect at the arrow marked "offset." Vertical, 4-inch diameter core; uphole is towards the top of the photo.

Figure B8.3 Deformation bands (white horizontal bands at the arrow heads) in a high-porosity, poorly bedded marine sandstone. The sandstone is white as indicated by the fresh break at the top of the core. The core surface has been stained gray by drilling mud except along the low-permeability/low-porosity deformation bands. Vertical, 4-inch diameter core; uphole is towards the top of the photo.

Figure B8.4 Two deformation bands appear as narrow white planes of comminuted rock in an oil-stained sandstone core. Oil staining is confined to the unaltered rock between the deformation bands, and reduced porosity adjacent to the planes of shear has minimized oil invasion for several millimeters on either side of the deformation bands. Vertical, 4-inch diameter core; uphole is towards the viewer.

Figure B8.5 Irregular deformation band shear planes composed of comminuted grains in sandstone, exposed on a slabbed core face. The shallow intersection angle between the plane of shear and the slab surface makes the shear planes appear to be more irregular than they are. This exposure also enhances the apparent width of the shear zone. Vertical, 4-inch diameter core; uphole is towards the top of the photo.

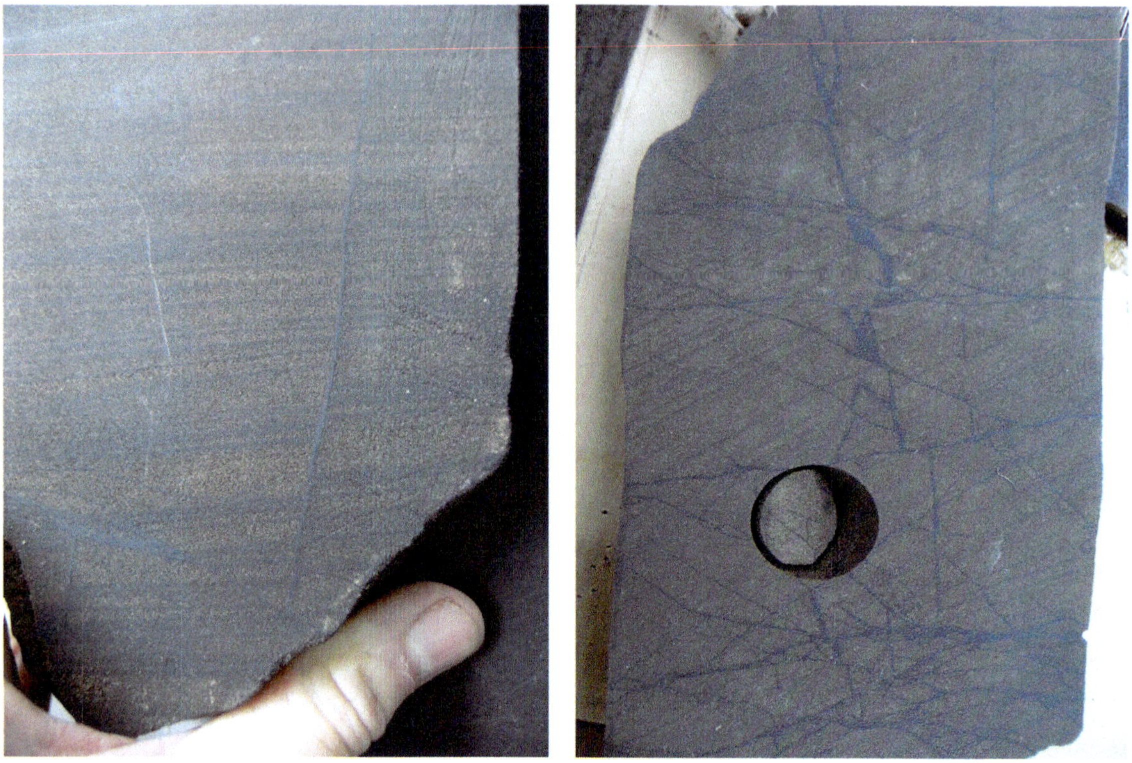

Figure B8.6 Many deformation bands are discrete structures that are easy to count and characterize (left). Others are part of more complex systems where bands join and split, making it difficult to count individual fractures. The orientations of fracture planes in this photo cannot be determined because they cannot be seen in three dimensions due to obscuration by bitumen. Bitumen was concentrated by capillary forces along the small grains forming the deformation bands, turning them black and obscuring the nature of the structure. Vertical, 4-inch diameter core; uphole is towards the top of the photos.

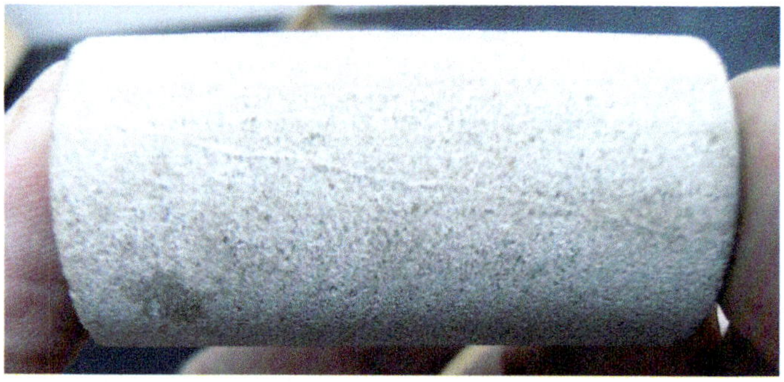

Figure B8.7 Oil and bitumen can be removed with toluene in the lab, revealing the underlying rock. This 1-inch plug was cut from the same oil-soaked core shown in the previous photo and has been flushed to reveal the white sandstone and the unmineralized deformation band.

References

Antonellini, M., and Aydin, A., 1994, Effect of faulting on fluid flow in porous sandstones: petrophysical properties. American Association of Petroleum Geologists Bulletin, 78(3), 355–377.

Aydin, A., and Johnson, A.M., 1978, Development of faults as zones of deformation bands and as slip surfaces in sandstone. Pure and Applied Geophysics, 116, 931–942.

Fossen, H., Schultz, R.A., Shipton, Z.K., and Mair, K., 2007, Deformation bands in sandstone: a review. Journal of the Geological Society, 164, 755–769.

Jamison, W.R., and Stearns, D.W., 1982, Tectonic deformation of Wingate Sandstone, Colorado National Monument. American Association of Petroleum Geologists Bulletin, 66, 2584–2608.

Nelson, R.A., 2002, Geologic Analysis of Naturally Fractured Reservoirs, 2nd edn. Boston: Gulf Professional Publishing.

Olsson, W.A., Lorenz, J.C., and Cooper, S.P., 2004, A mechanical model for multiply-oriented conjugate deformation bands. Journal of Structural Geology, 26, 325–338.

Rath, A., Exner, U., Tschegg, C., Grasemenn, B., Laner, R., and Draganits, E., 2011, Diagenetic control of deformation mechanisms in deformation bands in a carbonate grainstone. AAPG Bulletin, 95, 1369–1381.

B9

Faults

Core provides only small samples of faults. Slickenlines and slickencrysts can indicate the sense of motion along a fault (Figure B9.1), but the magnitude of offset on a cored fault can rarely be obtained. In fact, unless a fault is highly mineralized it may be represented only by wellbore washouts in an image log and only by rubble or missing intervals in a core. Unfortunately, the fact that an interval of core is missing is not always apparent, especially after a core has been processed and slabbed.

Smaller faults can be cored and recovered intact (Figures B9.2, B9.3), but incompetent fault breccia and gouge associated with larger faults may be ground up and lost, may fall out the bottom of the core barrel during retrieval, or may be recovered only as rubble unless they have been recemented after faulting (Figures B9.4, B9.5). Incompetent fault zones can jam in a core barrel, resulting in shortened core runs with no recovered evidence for faulting. Some cored faults are represented only by large pieces of mineralization, commonly with euhedral crystals indicating the presence of open void space along the fault (Figure B9.6), or by host-rock rubble displaying mineralized and/or slickenlined surfaces. The dominance of photos of intact cored faults in this section is only an indication of the fact that intact faults are more photogenic than fault rubble and does not indicate the ratio of intact to rubblized faults in cores.

Fault rubble can also be difficult to distinguish from rubble induced by coring processes such as the hammering needed to remove the last foot or so of core from the core catcher at the bottom of a core, or the hammering commonly used to remove jammed core pieces from an inner core barrel. This occurs less frequently now that split core barrel liners are in common usage, but there are still many archived cores in warehouses that were removed from the core barrel with the biggest available hammer. Rubble can also result from overfilling a core barrel (i.e., trying to put 32 feet of core in a 30 foot core barrel). Muddy bags of rubble, whether induced or related to faulting, are rarely examined during a core study and are often stored with the core butt boxes, not the slabs. Few service companies wash the drilling mud from such bagged rubble, and many bags of rubble have been thrown away. Nevertheless, bags of rubble often offer insights, and at least a few of the larger pieces should be cleaned up and examined for slickenlines or other indications of faulting when logging a core. If an image log was run in the hole, the intervals of the log correlative to core rubble zones should also be checked for possible faulting.

Even when recovered intact or nearly so, faults are easily broken up and significant information is lost during the handling and slabbing processes. The earlier a geologist can log a core, the more information the core will reveal about cored faults.

Atlas of Natural and Induced Fractures in Core, First Edition. John C. Lorenz and Scott P. Cooper.

Figure B9.1 Two views of a small fault with a 45° dip angle, captured in core cut from a marine shale. The component shear planes of the fault are mineralized with calcite and display slickencrysts indicating normal dip-slip offset. The shear planes that bound the fault zone are parallel to each other but there are multiple, intersecting shears between them. Bedding in the rest of the core is horizontal. Vertical 4-inch diameter core; uphole is towards the top of both photos.

Figure B9.2 A fault consisting of a discrete shear zone with evidence for repeated shear and calcite mineralization in the footwall, and an overlying zone of more diffuse shear, in core cut from a laminated marine silt-shale sequence. The 60° dip angle suggests that offset is normal dip-slip, but the fractography of the shear surface that would confirm it is not exposed. As demonstrated here, the degree of deformation is often greater in the hanging wall of a fracture than in the footwall. Vertical 2.5-inch diameter core; uphole is towards the top of the photo.

Figure B9.3 Two views of a fault in core cut from a laminated silt-shale sequence, displaying fault breccia and gouge on the hanging wall and lined on one side by calcite with evidence of repeated mineralization and shear. Vertical 2.5-inch diameter core; uphole is towards the top of the photo.

Figure B9.4 Two views of a cored, high-angle fault in a limestone-shale sequence. Left: the slab plane was cut nearly parallel to the irregular, generally high-angle plane of the fault (arrow). Right: the slab plane cuts across a concentration of breccia that fell to the bottom of the fault. The left photo shows the core butts pieced together, with the fault extending for several feet along the core. A layer of gray crystalline calcite mineralizes the fault surfaces and cements the fault breccia together but does not fully occlude the vuggy porosity along the fault. The nearly vertical dip angle of the fault suggests that this is a strike-slip fault, consistent with other strike-slip kinematic indicators in the core. Slabbing has degraded the integrity of the fault and some of the fault characteristics, but it has also opened up and revealed some of the three-dimensional geometry of the fault. Vertical, 4-inch diameter core; uphole is towards the top of both photos.

Figure B9.5 A cored, complex, incompletely mineralized, high-angle, strike-slip fault in a dolomite core. The fault extends for several feet along the vertical core and is locally represented only by rubble pieces displaying calcite-mineralized fracture faces. Locally, the recemented fault breccia forms a matrix with centimeter-scale voids lined with calcite crystals. Vertical, 4-inch diameter core; uphole is towards the top of the photo.

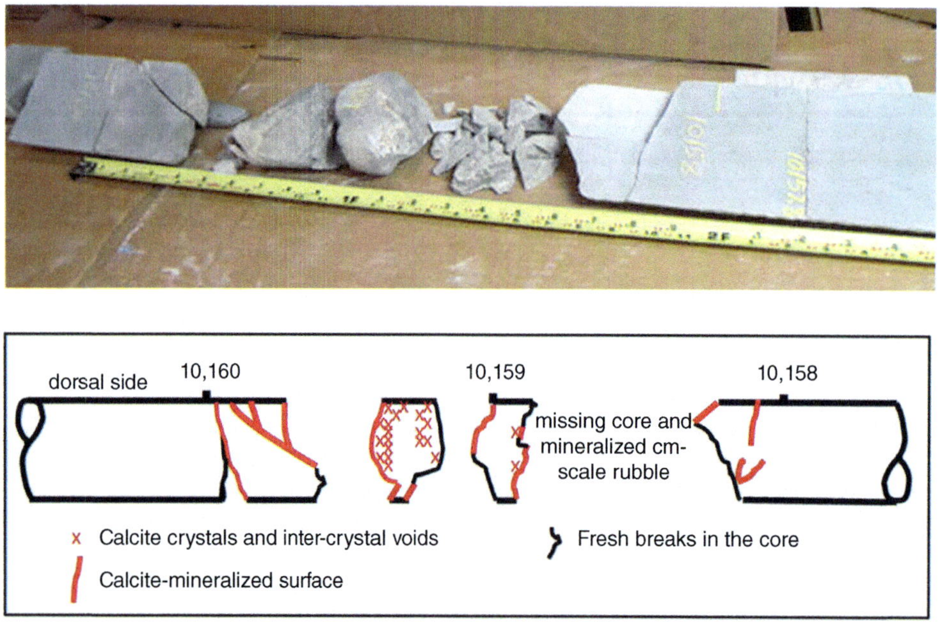

Figure B9.6 Many faults in cores are represented by broken and missing rock, especially after the core has been slabbed and souvenirs of the crystalline mineralization have found their way to a manager's desk. This fault in the butts of a horizontal sandstone core, examined after sampling and slabbing, was removed from the core box and pieced together as much as possible. Depth markings on the core remnants indicate the amount and location of missing core. The sketch below the photo indicates the width of the sheared interval as well as the degree of mineralization. Horizontal 4-inch diameter core; uphole is towards the right of the photo, stratigraphic up is towards the top of the photo.

Section C

Other Types of Natural Fractures

C1

Introduction

Extension and shear fractures as previously described are the most common types of natural fractures found in cored hydrocarbon reservoirs, but there are other, less common types of natural fractures in cores. Some of these fracture types are unique examples of extension or shear fracturing (i.e., ptygmatically folded fractures, compaction fractures, and veins) that formed as synsedimentary features in specific sedimentary settings, and which are not related to structure or tectonics. One type of fracturing is related to overpressuring and expulsion of fluids from the system, but this is not a common fracture type and has very specific, diagnostic characteristics; many fractures have been misidentified as expulsion fractures for no other reason than the core contains fractures.

Some of the structures illustrated in this section are not fractures per se, but rather they are the alteration products of diagenesis and dissolution along fractures. Dissolution-enhanced fractures can be enlarged to become poorly defined fissures. Continued dissolution along a system of fissures can produce caves and karsted rock, and the caves may become filled with karst-related breccia with non-systematic fractures defining the breccia clasts. The breccia in turn can become cemented into a conglomeratic rock that can itself become naturally fractured.

Tectonic and structural stresses produced some of the other fracture types illustrated in this section, including submillimeter-scale microfractures and stylolites. Microfractures are too small to be an important part of this atlas of core-scale features, but they have been extensively studied and have been suggested to be an important part of the permeability system in some reservoirs, so a few examples are included. Stylolites are planar features formed by pressure solution; like deformation bands, they are not fractures in the usually accepted sense of the term, but the tension fractures that are commonly associated with stylolites are, and both are illustrated here.

The rock cut by some cores is so completely fractured that it has a shattered texture with few characteristics that help to define the origin of the shatter zone. Other cored structures consist of several logically relatable substructures including fractures that together form a consistent, easily interpreted, core-scale geomechanical system.

Atlas of Natural and Induced Fractures in Core, First Edition. John C. Lorenz and Scott P. Cooper.

C2

Microfractures

The term "microfracture" typically refers to fractures which can only be viewed easily with a microscope, and as such they do not fit the scope of this book which is to illustrate core-scale fracturing. Nevertheless, microfractures are worth a short discussion as the permeability and some of the porosity in certain reservoirs are deemed to depend on these features.

Mineralized microfractures (Figure C2.1) have been documented within, across, and between the component grains of many cored formations (Anders et al., 2014; Zeng, 2010; Zeng and Li, 2009). Zeng also illustrates unmineralized microfractures, while Anders et al. suggest that most natural microfractures should be mineralized. In fact, there have been arguments as to whether *un*mineralized microfractures in cores are natural or whether they are the product of stress relaxation after the core has been cut free from the in situ confining stresses. Natural, tectonically created, unmineralized microfractures may be present in some reservoirs but, as with all fractures, they should be assessed carefully and critically; the technique of anelastic strain recovery was developed specifically to measure in situ stress orientations and magnitudes from the development of the stress-release microfractures that form in cored rock during the first few hours after it has been cut free from a formation at depth (e.g., Teufel, 1983; Warpinski et al., 1993).

Figure C2.1 Calcite-mineralized microfractures in a tight, quartz sandstone. The fractures cut across some grains and along the edges of others. Blue epoxy glue highlights narrow, unmineralized extensions of these fractures of unknown origin.

Perhaps the best way to distinguish natural from induced unmineralized microfractures would be to document their orientations relative to the in situ stress system using oriented thin sections. Induced, relaxation microfractures should strike predominantly perpendicular to the present-day maximum horizontal in situ compressive stress, whereas tectonic microfractures should strike parallel to that stress as long as the stress orientations have not changed since fracturing. This can be determined even in unoriented cores if the thin sections can be orientated relative to the strikes of stress-controlled induced petal or centerline fractures described in Part 2 of this atlas. Some authors suggest that almost all natural microfractures are mineralized, others suggest that relaxation microfractures occur primarily between grains whereas tectonic microfractures occur more commonly within grains (S. Brown, personal communication, 2005).

Some small fractures are too large to be considered microfractures yet are too small to be logged individually or to be used as discrete inputs into reservoir models (Figure C2.2). When logging, rather than recording individual fractures, the fracture fabric can be estimated by noting whether or not the fractures are parallel to each other, recording the maximum, minimum, and estimated average dimensions for width, height, length, and spacing, and by assessing the approximate number of fractures per foot of core as well as their distributions with respect to lithology.

Atlas of Natural and Induced Fractures in Core, First Edition. John C. Lorenz and Scott P. Cooper.

Figure C2.2 Short (both laterally and vertically) unmineralized extension fractures in a limestone core are highlighted by the oil that is bleeding to the core surfaces. These do not fit a strict definition of "microfractures" in that they can be seen without the aid of a microscope, but they are too small to have distinct signatures on an image log. Similar systematic fractures are present along most of this core. Vertical, 4-inch diameter core; uphole is towards the top.

References

Anders, M.H., Laubach, S.E., and Scholz, C.H., 2014, Microfractures: a review. Journal of Structural Geology, 69, Part B, 377–394.

Teufel, L.W., 1983, Determination of In-Situ Stress from Anelastic Strain Recovery Measurements of Oriented Core SPE-11649-MS, Society of Petroleum Engineers, SPE/DOE Low Permeability Gas Reservoirs Symposium, 14–16 March, Denver, Colorado.

Warpinski, N.R., Teufel, L.W., Lorenz, J.C., and Holcomb, D.J., 1993, Core-Based Stress Measurements: A Guide to Their Application. Gas Research Institute Topical Report GRI-93/0270. Available from the Gas Technology Institute or through the US Government Office of Science and Technology Information: www.osti.gov.

Zeng, L., 2010, Microfracturing in the Upper Triassic Sichuan Basin tight-gas sandstones: tectonic, overpressure, and diagenetic origins. AAPG Bulletin, 94, 1811–1825.

Zeng, L., and Li, X-Y., 2009, Fractures in sandstone reservoirs with ultra-low permeability: a case study of the Upper Triassic Yanchang Formation in the Ordos Basin, China. AAPG Bulletin, 93, 461–477.

C3

Ptygmatically Folded Fractures

Early-formed, ptygmatically folded, high-angle extension fractures are common in cores cut from many marine resource-play shales and some muddy limestones. In some formations, these fractures are confined to the more calcareous or siliceous layers (Figures C3.1–C3.3), with their short heights being related to bed thickness in thinly bedded formations. In other formations, they are preferentially developed in the muddier lithologies, and they can be several feet tall in formations where bedding is poorly defined or consists of successive beds with similar compositions and mechanical properties (Figure C3.4). Fracture intensity varies significantly, from scattered small fractures to systems consisting of 6–10 intensely fractured beds in each foot of core.

Ptygmatically folded fractures tend to be short, narrow, and tightly mineralized, and are therefore difficult to distinguish on an image log. However, the conductive pyrite mineralization in some of these fracture systems can increase the probability of a log signature even if the small individual fractures cannot be distinguished.

These fractures formed initially as unfolded structures, but became folded and otherwise reduced in height to accommodate vertical shortening as the host strata were compacted during burial (Figures C3.5, C3.6). Where a fracture extends across multiple lithologies, it commonly shows greater degrees of folding in the more compacted strata (Figure C3.7). Folding is the most common strain accommodation structure, but shortening was also accommodated locally by vertical overlaps of the fracture plane mineralization that resemble millimeter-scale vertical thrust belts (see Figure C3.6), and by wedging, where the mineralization has been driven into itself along the fracture plane and resembles the wedges that hold hammer heads to wooden handles (Figure C3.8).

The bedding planes where these fractures terminate are commonly bent upward and downward to form ridges along the fractures, the mineralization apparently having formed stiff struts that minimized compaction in the immediate vicinity of the fracture (Figure C3.9). Elsewhere, the stiff mineralized fracture planes were pushed a short distance into the under- and overlying strata as the rock compacted. Some fractures were smeared out laterally as the sedimentary column was sheared sideways during compaction.

These fractures are commonly filled with a different kind of mineralization than later fractures in the strata, often including pyrite suggestive of early diagenesis.

Ptygmatically folded fractures are typically fully mineralized, and since the opposing fracture walls interfinger along the folds and are cemented together by mineralization, the fracture surfaces are rarely exposed. In rare cases where the surfaces are visible, they are regularly indented by the small folds and resemble the texture of corduroy cloth (Figures C3.10, C3.11). Unlike the plume structure of extension fractures or the slickenlines of shear fractures which record mode of origin, the small subparallel ridges ornamenting the surfaces of ptygmatically folded fractures record postfracture compaction.

These structures are sometimes interpreted to be mud cracks or synaerisis cracks, but they pinch out top and bottom rather than opening at a sediment–water or sediment–air interface. Bishop et al. (2006) relate these structures to early diagenesis and gas generation in poorly lithified sediment that was not deeply buried at the time of fracturing, and suggest that they are similar in form to molar tooth structures that have been described from Precambrian shales.

Incompletely lithified, muddy strata are not commonly thought of as fracture-prone media, but materials that are ductile at room conditions can become brittle and susceptible to fracture under conditions of high pore pressure. Even rubber can fracture under the right conditions.

Ptygmatically folded fractures can be well enough developed in some reservoirs to possibly be of significance to reservoir plumbing, especially where they formed as intersecting fracture sets. Since they are typically developed in strata with permeability measured in nanodarcys, even low permeabilities along mineralized but intersecting fractures may be significant if the fractures are closely spaced and have a degree of intersection.

Atlas of Natural and Induced Fractures in Core, First Edition. John C. Lorenz and Scott P. Cooper.

Taller, minimally folded ptygmatic fractures can resemble non-folded postlithification extension fractures, being differentiated where the two fracture types occur together by the degree of folding (Figure C3.12), commonly by slightly different strikes, and by intersections showing that the ptygmatic fractures are older. Tall ptygmatic fractures grew both upward and downward, as indicated by hooking relationships between the top of one fracture and the base of an overlying fracture (Figure C3.13).

Most ptygmatically folded fractures occur as sets of parallel planes in any given bed, and the fractures in successive beds are commonly parallel or subparallel to each other, suggesting a consistent stress anisotropy in the system even though it was poorly lithified at the time of fracturing. A shallow depositional slope such as the quarter-degree slope of the US Gulf Coast sea floor would provide the necessary stress differential.

Some ptygmatically folded fractures occur as two oblique sets of parallel fractures in successive beds or even as superimposed oblique sets in the same bed. We have observed two distinct strikes with a 30° difference in one core (Figure C3.14). Elsewhere the fractures form less systematic, intersecting systems (Figure C3.15).

Figure C3.1 Numerous short, ptygmatically folded fractures concentrated in the calcareous layers of a marine shale core. Some of the fractures have visible but small-scale remnant fracture porosity despite being mineralized with calcite. Slabs from a vertical, 4-inch diameter core; uphole is towards the top of the photo.

Figure C3.2 Two generations (wide and narrow) of calcite-mineralized, high-angle, ptygmatically folded extension fractures are concentrated in the more calcareous layers of a marine shale. Bedding boundaries are tented upward and pushed downward near the larger fractures, indicating that the planes of fracture mineralization formed stiff units in the package and resisted compaction. Slabs from a vertical, 4-inch diameter core; uphole is towards the top of the photo.

Figure C3.3 Two quartz-filled fracture planes within a variably siliceous bed that is bounded by darker, thinner, muddy shales. The fracture planes have been shortened vertically, leading to ptygmatic folding in the less siliceous lithology near the upper and lower ends of the fractures, and shear offset and repetition in the more siliceous zone in the middle of the bed.

Figure C3.4 Some ptygmatically folded extension fractures extend vertically for several feet along a core. In addition to the small-scale folding, they are typically less planar than the associated extension fractures that formed after lithification. Slabs of vertical, 4-inch diameter core; uphole is towards the top of the photo.

Figure C3.5 Several scattered, irregular, ptygmatically folded fractures in a muddy limestone core. Bedding is poorly developed, and the fractures are not restricted to specific beds. Slab of a vertical, 4-inch diameter core; uphole is towards the top of the photo.

Figure C3.6 Two folding styles are illustrated by this photo of a calcite-mineralized fracture in a marine shale. The upper part of the fracture did not become shortened, suggesting that there was little compaction in this more calcareous interval of the host rock. The middle part of the fracture shortened by overlap of stiff struts formed by the mineralization that filled the fracture aperture. The lower part of the fracture was folded as the more clay-rich host zone compacted. Vertical 4-inch core; uphole is towards the top of the photo.

Figure C3.7 Two views of an early-formed, high-angle extension fracture that is not strata-bound, where the degree of ptygmatic folding varies depending on the degree of compaction of the host lithology. In the fine-grained, massive limestone beds, folding is minimal and this early fracture is nearly indistinguishable from later, more planar fractures. In contrast, folding of the same fracture is well developed where it crosses muddy interbeds. Butt (left) and slab (right) of vertical, 4-inch diameter core; uphole is towards the top of both photos.

Figure C3.8 Three views of a short but laterally extensive, pyrite-mineralized, ptygmatically folded fracture in a muddy shale. The top photo shows a vertical section across the fracture on the slab face of the core; the middle photo shows a two-dimensional view of the fracture including the horizontal plane where the core has split along the bedding plane indicated by the arrow in the first photo; the bottom photo shows the top surface of that bedding plane, illustrating the laterally extensive and linear pattern common to these fractures in the horizontal plane, as well as the multi-stranded pattern created by the overlapping of segments of the mineralized fracture plane during compaction.

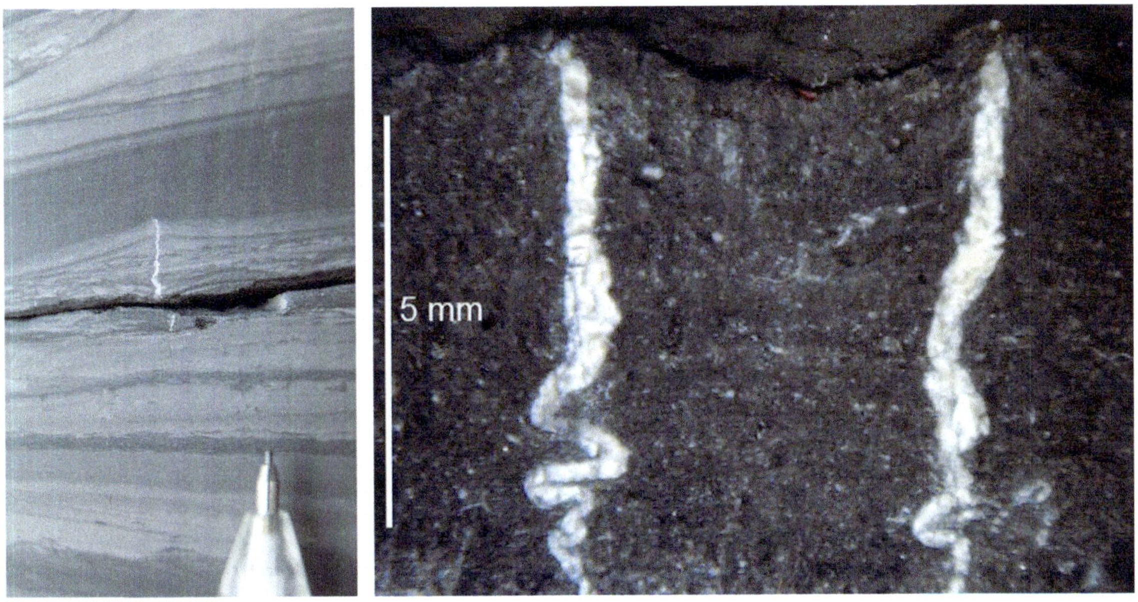

Figure C3.9 Left: a small, isolated, ptygmatically folded fracture in a calcareous layer in core from a marine shale inhibited compaction in the sediment adjacent to the fracture, tenting the bedding locally. Where exposed by core breakage, such tented bedding forms linear ridges across the bedding plane. Right: two parallel, calcite-filled fractures in a marine shale hold up ridges along in the bedding plane. Slabs of vertical, 4-inch diameter core; uphole is towards the top of both photos.

Figure C3.10 The corduroy-like surfaces of two ptygmatically folded fractures in core from two different calcareous shale formations are marked by small, subhorizontal ridges which are the impressions left in the poorly lithified sediment by the small, linear, horizontal ptygmatic folds during compaction. The surface of the bottom fracture is obscured by calcite mineralization. Vertical 3-inch cores; uphole is towards the top of both photos.

Figure C3.11 The faces of some ptygmatically folded fractures are less regular because the fractures are too tightly folded to leave ridges, and the broken surface consists in part of the fracture plane and in part of the broken host rock. Vertical 4-inch core; uphole is towards the top of the photo.

Figure C3.12 An early-formed, ptygmatically folded, vertical extension fracture (red arrow) and a later-formed, more planar, high-angle extension fracture (blue arrow) in a calcareous marine shale. A saw cut across the core shows that the two fractures intersect at a 25° angle, and that the planar fracture changes strike, being deflected slightly where it crosses the older, ptygmatically folded fracture. Vertical, 3-inch diameter core; uphole is towards the top of the photo.

Figure C3.13 Hooking of the tips of two ptygmatically folded fractures towards each other in a massive fine-grained limestone suggests that they grew upward and downward and that the propagation and termination of each fracture were influenced by the proximity of the other. Slabs from vertical, 4-inch diameter core; uphole is towards the top of the photo.

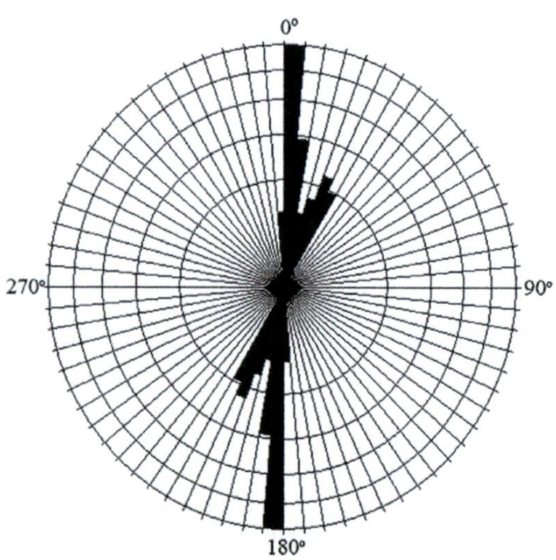

Figure C3.14 Strikes of two sets of early high-angle ptygmatic extension fractures relative to the orientation reference provided by coring-induced petal fractures. Core is from a calcareous marine shale (n = 237) and data are from CT scans and physical core examination. Note that 0° is not north but denotes the strike of the induced reference fractures.

Figure C3.15 Some ptygmatically folded fractures have less systematic strikes as indicated by the impression of fractures on this bedding plane. The core is oriented ("PSL" indicates the location of the Principle Scribe Line, an orientation groove cut onto the core surface); the question mark indicates that the orientation survey was of doubtful reliability. Fracture strikes of 25°, 40°, 90°, and 140° were calculated from the orientation information, and the marked petal fracture, exposed on the side of the core, was calculated to have a NW-SE strike. The circled dot indicates that the view is downhole. Vertical, 4-inch diameter core; uphole is towards the viewer.

Reference

Bishop, J.W., Sumner, D.Y., and Huerta, N.J., 2006, Molar tooth structures of the Neoarchean Monteville Formation, Transvaal Supergroup, South Africa. II: a wave-induced fluid flow model. Sedimentology, 53, 1069–1082.

C4

Fissures

As used here, the term "fissure" refers to fracture-like structures with variable characteristics. "Fissures" are irregular but roughly planar features with rough, non-uniform widths suggestive of dissolution. They are commonly filled with poorly sorted exotic materials (Figure C4.1), probably introduced into the open slots at an overlying surface exposed to weathering and dissolution. Some are filled with locally derived materials from the host rock which are of more uniform composition (Figure C4.2).

Filled fissures can be difficult to distinguish from the exotic materials injected into a rock from an unconsolidated adjacent source bed during deposition or from fault breccias. Fissures commonly form during the early stages of karstification along dissolution-enhanced fractures. They can form networks following the original fracture system, where the only evidence remaining for the original fractures after dissolution may be the large-scale geometry of the fissure network, which is difficult to document in the small samplings provided by core.

A well-developed fissure system in a reservoir can significantly enhance reservoir quality depending on the nature and permeability of the filling material as well as on the spacing and vertical extent of the fissures. Fissures should show in image logs but since they have irregular geometries they would be poorly defined and difficult to interpret.

Figure C4.1 Three views of a fissure formed in a vuggy limestone and filled with exotic calcareous debris. The irregular width of the structure suggests it was formed or at least enhanced by dissolution. The fissure is truncated by and therefore predates a stylolite. Slabs of vertical, 4-inch diameter core; uphole is towards the top of the left and lower right photos, and towards the viewer in the upper right photo.

Atlas of Natural and Induced Fractures in Core, First Edition. John C. Lorenz and Scott P. Cooper.

Figure C4.2 A fissure developed in a chalky limestone and filled with angular blocks of poorly sorted, locally derived breccia. The yellow-orange material is calcite precipitated into void spaces. There is no evidence for shear, indicating that this is not a fault breccia. A quarter slab of a vertical, 4-inch diameter core; uphole is towards the top of the photo.

C5

Veins

As used here, the term "veins" covers a variety of features of uncertain origin. Some veins in fine-grained carbonates, particularly in dolomites, may be diagenetic in origin. Veins are typically completely filled with an amorphous, non-crystalline mineralization, commonly of a composition similar to that of the host rock (Figure C5.1), suggesting formation early in the sedimentation, burial, and diagenetic history of the rock. Many veins have embayed walls supporting an interpretation of dissolution.

Some veins are short and strata bound (Figure C5.2), others are taller and cut across several beds. Veins may be short and wide, tapering abruptly to upper and lower terminations, indicating formation in relatively ductile or incompletely lithified strata. Others terminate at bedding planes, suggesting syn-sedimentary but post vein formation dissolution along a bedding surface (Figure C5.3).

Veins should have little effect on a reservoir. They are typically occluded by material with the same mineralogy as the host rock and may not have a geophysical signature in an image log.

Figure C5.1 Large, calcite-filled veins in limestones. The vein at the left tapers and pinches out at the top and bottom; the vein at the right pinches out downward but terminates abruptly at the top at a bedding plane. Slabs of vertical 4-inch diameter core; uphole is towards the top of both photos.

Atlas of Natural and Induced Fractures in Core, First Edition. John C. Lorenz and Scott P. Cooper.

Figure C5.2 Two views of small, parallel, high-angle, dolomite-filled veins in a fine-grained dolomite. Butts of vertical, 4-inch diameter core; uphole is towards the top of the left photo and towards the viewer in the right photo.

Figure C5.3 A wide, calcite-filled vein in a limestone terminates at an irregular surface located between the arrows. This bedding plane is a few centimeters higher in the core on the left side of the vein, suggesting that the plane is a dissolution surface. The vein was truncated by the same dissolution, having formed prior to dissolution. Vertical 4-inch diameter core; uphole is towards the top of the photo.

C6

Expulsion Structures

Expulsion structures are not uncommon in the rock record, ranging from igneous intrusions to sedimentary injectites such as the mud mounds at the Mississippi River delta. Such expulsion/injection structures require a high-pressure external source and a low-pressure destination. Pressure differentials capable of initiating expulsion can be generated when sediments are buried before being completely dewatered (undercompaction), and/or when the elevated temperature and pressure associated with burial cause organic maturation, resulting in excess fluid volume and high pore pressures in some parts of a formation but not others (see Osborne and Swarbrick, 1997). Expulsion commonly occurs in cycles; the pressure differences that are episodically equalized by expulsion and injection typically build back up again to exceed a threshold pressure, initiating a new round of expulsion/injection.

Caution should be used in inferring expulsion as the origin of fractures; expulsion may be attractive conceptually, but the fact that something *can* happen does not mean that it did, and in fact, only a few fracture types have characteristics that are compatible with this mechanism.

If a stress differential exists in a formation, expulsion and injection should form parallel structures that are irregularly vertical or horizontal sheets (Figures C6.1–C6.3). Expulsions of high-pressure fluids commonly carry local sediment and rock bits with them, and one common characteristic of expulsion structures is the presence of exotic materials and maybe the remnants of exotic fluids within the resulting injection structure (see Figure C6.1). Expulsion also commonly disrupts bedding in the recipient formation, and since they are typically episodic, expulsion fractures frequently occur in bundles (Figures C6.1–C6.4).

Filled expulsion fractures can resemble fissures or fault gouge, but the absence of evidence for dissolution or shear helps to distinguish them in cores. Because they are typically filled, they have little effect on most reservoirs, and they are hard to diagnose in image logs because they are irregular.

Figure C6.1 Two views of a multi-stranded expulsion fracture system in a dolomite. Left: core showing the multiple strands produced by repeated injection with disrupted materials injected by the expulsion. Right: the top of the same core piece, illustrating the irregular planarity of the structure. Vertical 4-inch diameter core; uphole is towards the top of the left photo and towards the viewer in the right photo.

Atlas of Natural and Induced Fractures in Core, First Edition. John C. Lorenz and Scott P. Cooper.

Figure C6.2 An irregularly planar dewatering structure in muddy, silty fluvial deposits, created by the expulsion of muddy water from an overpressured, underconsolidated, underlying mudstone. Burial increased the fluid pressure in the undercompacted mudstone, and as the pressure reached a threshold, an expulsion pathway was formed in the siltstone cap, reducing the fluid pressure in the mudstone and temporarily terminating the process. The pathway was then plugged by calcite, and the cycle was repeated as burial and compaction continued. The planarity of the structure (normal to the plane of the photo) suggests semi-lithification and the presence of a stress anisotropy in the formation at the time of expulsion. Vertical, 4-inch diameter core; uphole is towards the top of the photo.

Figure C6.3 An irregular, multi-stranded, calcite-mineralized structure related to repeated dewatering in a marine shale. The structure is roughly planar in plan view, suggesting that a stress anisotropy existed in the poorly lithified strata at the time of expulsion. The top core piece has been turned to expose the fracture in the horizontal plane. Vertical, 4-inch diameter core; uphole is towards the top of the photo.

Figure C6.4 Expulsion fractures in a diatomite. Upper left: multiple strands record multiple expulsion events, and the irregularity of the dips suggests poor lithification and postinjection soft-sediment deformation. Upper right: the face of one strand showing the injected material (hydrocarbons and clays) lining the fractures. Bottom: expulsion pathways were diverted around a syn-sedimentary concretion in the poorly lithified sediment. Vertical 4-inch diameter core; uphole is towards the top of all three photos.

Reference

Osborne, M.J., and Swarbrick, R.E., 1997, Mechanisms for generating overpressure in sedimentary basins: a re-evaluation. AAPG Bulletin, 81, 1023–1041.

C7

Syn-Sedimentary Fractures

A variety of structures fall into the basket category of syn-sedimentary fractures. Most of them are shears associated with pre- or syn-lithification compaction, but small-scale injections and extension fractures including mud cracks can also be lumped into this category. As always, as much of the full three-dimensional characteristics of these structures as possible, including fracture-face fractography, should be assessed before making an interpretation; the limited two-dimensional exposures offered by core slab surfaces or photographs often provide insufficient information.

Compaction shears are common in cores cut from muddy strata deposited in both marine and non-marine environments, but they typically have little expression on a rough outer core surface. Their presence may not be apparent until processing and handling when the core breaks along weakness planes provided by these slickenlined surfaces, and when the fracture count seems to double during each processing step. Compaction shears in muddy strata typically have glossy or locally scaly surfaces and intermediate dip angles. Slickenlines usually indicate shear parallel to the dip azimuth (Figure C7.1), and bedding adjacent to the shears may exhibit soft sediment deformation (Figure C7.2). The surfaces are not perfectly planar, suggesting shear prior to complete lithification when the sediment was ductile enough to accommodate shear on surfaces with compound curves without producing voids (Figure C7.3). Small chatter marks transverse to the slickenlines may be present, indicating that the rock was semi-lithified at the time of shear. Intersecting fracture planes and planes that intersect when projected beyond the core are common (Figures C7.4, C7.5)

A study of compaction shears in oriented core from a non-marine formation (Finley and Lorenz, 1988) shows that they have random strikes (Figure C7.6), supporting an early origin prior to the development of a horizontal stress anisotropy in the rock. Irregular syn-sedimentary shears are also commonly associated with differential compaction around early-formed nodules and below rapidly deposited sedimentary loads (Figures C7.7, C7.8). The ideal, Andersonian normal-displacement shear dip angle would be 60°, but dip angles decrease during compaction in semi- to unlithified sediment and compaction shears can have very low dip angles (see Figures C7.7, C7.8).

Most compaction fractures are unmineralized, but thin layers of calcite on some shears show that they were permeable at some point in their history. Nevertheless, different initial pressures have been measured in sandstone reservoirs separated by muddy intervals tens of feet thick and containing numerous compaction shears; this and observations of the lack of pressure communication between these reservoirs during production suggest that compaction shears are not significant permeability pathways even where they are well developed.

These unmineralized, closed fractures are difficult to observe in image logs but this point may be moot given that they have little influence on the reservoir. Nevertheless, it is important to recognize these structures and to differentiate them from larger scale shear fractures that do have significant influence on a reservoir permeability system.

Syn-sedimentary breccias can also be formed at erosion/exposure surfaces during a hiatus in deposition (Figure C7.9), typically in non-marine strata. They can be differentiated from fault breccias by a depositional rather than sheared contact with the overlying strata, and sometimes by a vertical gradation in clast size.

Syn-sedimentary desiccation cracks are also common in some cores, recognizable as short, wide, downward-tapering wedges with irregular strikes (Figure C7.10). Gradual rather than abrupt tapering and less than planar surfaces are both characteristic of fractures formed in semi-lithified strata (Figure C7.11) (see also section C5, Veins). Other syn-sedimentary fracture systems consist of high-angle extension fractures in thin brittle units combined with compatible inclined shears in the more ductile adjacent layers (Figure C7.12).

Atlas of Natural and Induced Fractures in Core, First Edition. John C. Lorenz and Scott P. Cooper.

Figure C7.1 Compaction shears with intermediate dip angles in muddy strata, showing irregular, roughly planar, slickenlined surfaces. Marine shale (left) and non-marine mudstone (right). The irregularly curved planes of these shears would accommodate shear only if the strata were poorly lithified and relatively ductile at the time of shearing. Vertical 4-inch (left) and 3-inch (right) core; uphole is towards the top of both photos.

Figure C7.2 The irregular plane of this faintly slickenlined, unmineralized, intermediate-angle fracture in horizontal core from a marine sandstone, as well as the associated soft sediment deformation, suggest that this shear formed prior to lithification. Slabs from a horizontal, 4-inch diameter core; uphole is towards the right, stratigraphic up is towards the top of the upper photo and obliquely into and towards the top of the lower photo.

Figure C7.3 Bedding shows the magnitude of offset on this calcite-mineralized, intermediate-angle shear plane in a marine mudstone. Small-scale bedding deformation at the shear contact as well as the slightly convex-upward curvature of the shear plane indicate that it formed in semi-lithified sediment. Slab from 3-inch diameter vertical core; uphole is towards the top of the photo.

Figure C7.4 Left: core containing three compaction shears with intersecting, slickenlined planes in a muddy marine shale. Right: close-up of the lower two shear planes shown in the left photo. Vertical, 4-inch core; uphole is towards the top of both photos.

Figure C7.5 Two views of a cone-shaped compaction feature composed of three shear planes with intersecting strikes and shear in multiple directions. Simultaneous shear in consolidated rock would cause unsolvable volume problems, so the shears must have formed in soft sediment and were probably sequential. Left photograph is a top-down view and right photograph is a side-view. Vertical, 3-inch core; uphole is towards the viewer in the left photo and towards the top of the right photo.

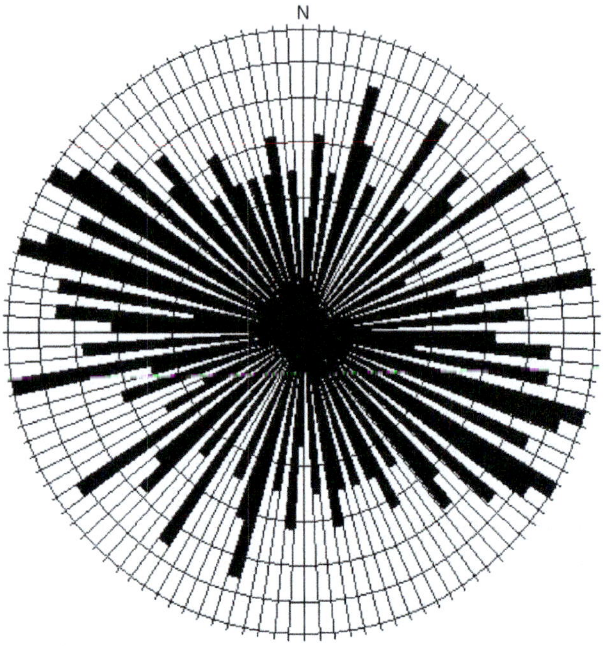

Figure C7.6 The irregular strikes of compaction shears in mudstones from the non-marine Mesaverde Formation in Colorado indicate that they formed in an isotropic horizontal stress regime and were not tectonically controlled. From Finley and Lorenz (1988). Three-degree orientation bins, n = 128.

Figure C7.7 Slickenlined, curved, and glossy shear planes formed by compaction around pyrite nodules in a marine shale. Vertical 4-inch diameter core; uphole is towards the viewer.

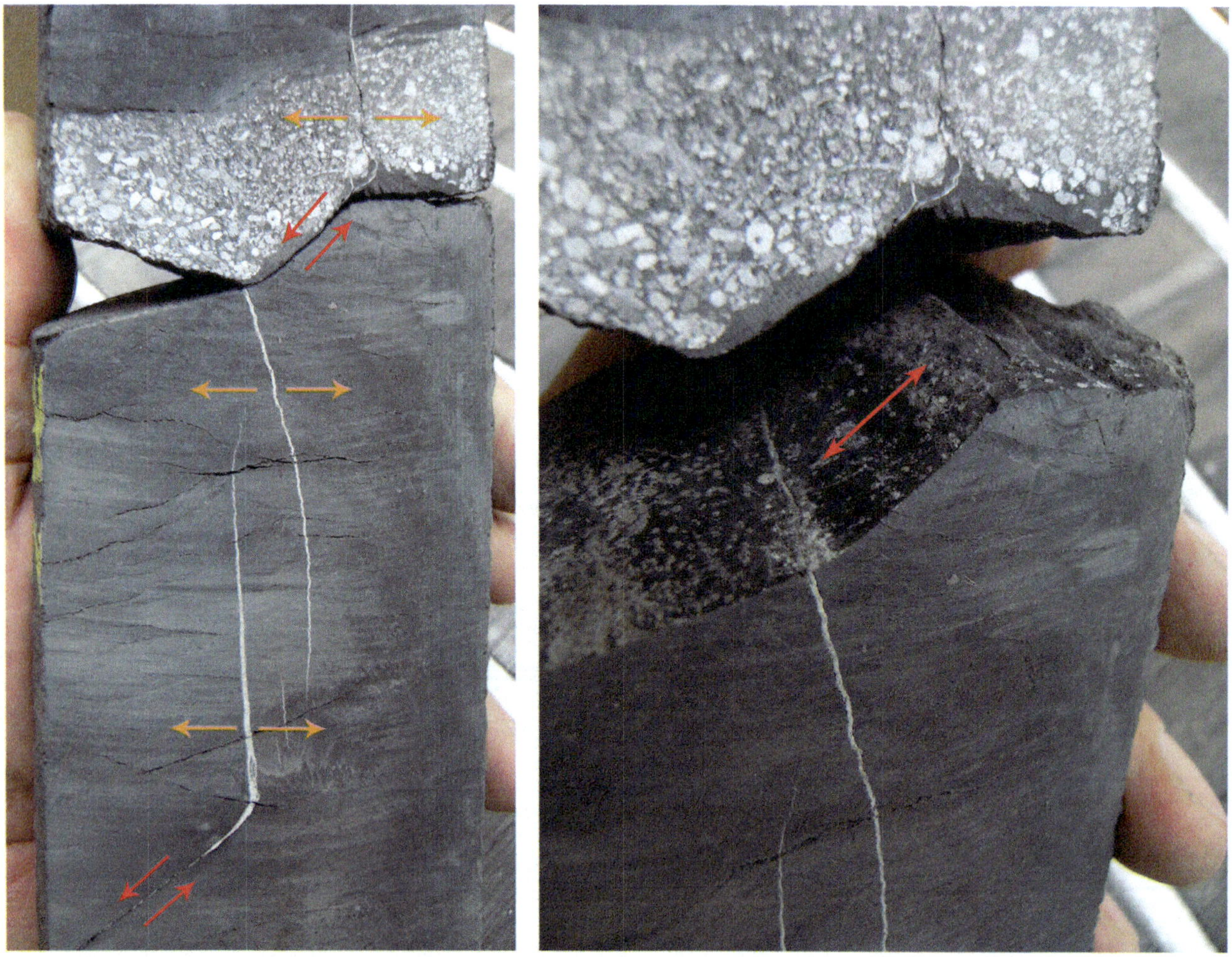

Figure C7.8 Compaction below this rapidly deposited carbonate debris flow in a marine shale produced an inclined, slickenlined shear plane (red arrow in the right photo) and associated extension fractures (orange arrows) in the underlying, semi-lithified mudstone. Vertical 4-inch diameter core; uphole is towards the top of both photos.

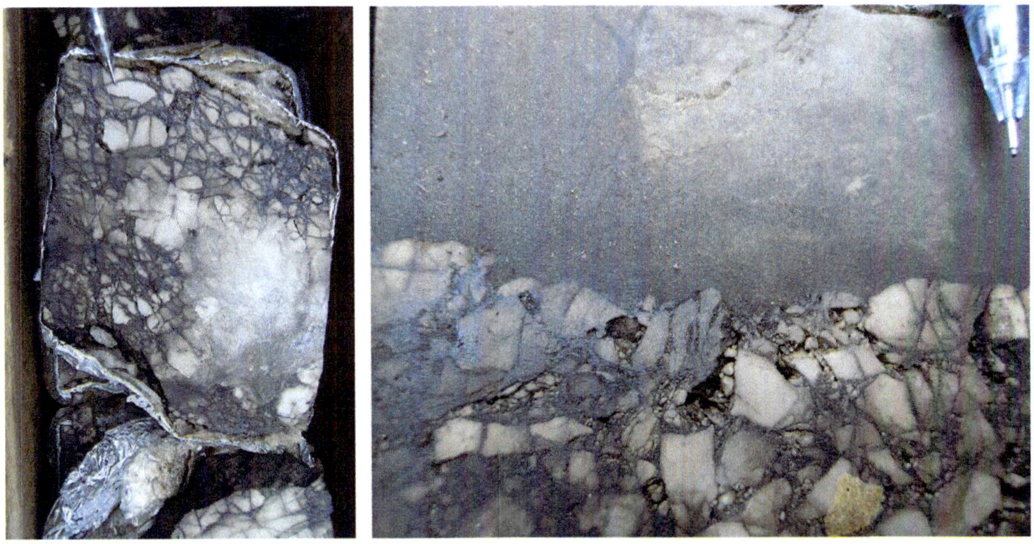

Figure C7.9 Context is important but often hard to obtain. What looks like a dolomitic fault breccia in the core slab held together by aluminum foil (left) is in fact an oil-stained sedimentary breccia as indicated by the depositional contact (right) between the breccia and the overlying oil-stained eolian sandstone in the core butt. Vertical 4-inch diameter core; uphole is towards the top of both photos.

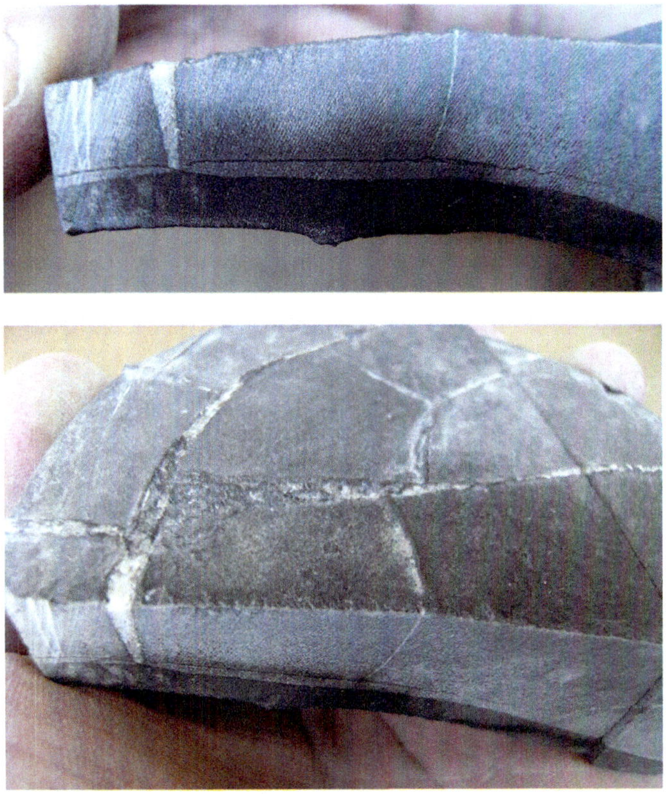

Figure C7.10 Two views of downward-wedging, short, wide, calcite-filled fractures formed as desiccation cracks in a nominally marine mudstone. The bedding-plane surface (bottom photo) shows a three-dimensional pattern diagnostic of mud cracks, and similar bedding surfaces in the system confirm that this is a characteristic pattern in the formation rather than an anomaly. Vertical, 4-inch diameter core; uphole is towards the top of the upper photo, and obliquely out of the plane of the photo and towards the viewer in the lower photo.

Figure C7.11 Gradually tapering terminations (left) and irregular planes with unexpected branches (right) can characterize extension fractures formed prior to complete lithification. These features distinguish this fracture suite from the later, more planar extension fractures found in the same core but which formed after lithification, and which are characterized by uniform widths and planar surfaces. Slabs from vertical, 3-inch diameter core; uphole is towards the top of both photos.

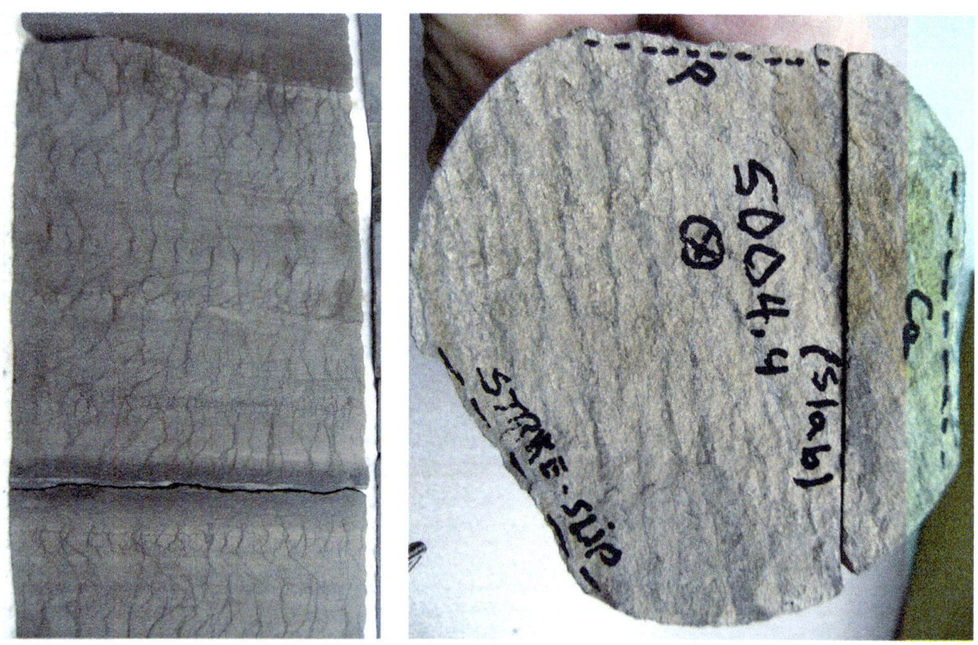

Figure C7.12 Two views of a syn-sedimentary extension/shear fracture system in a marine diatomite. Left: the system consists of high-angle extension fractures in the more competent units and intermediate-angle shear fractures in the less competent units. Right: submillimeter shear is documented by the low-relief offsets on the bedding plane cut by the shear fractures. These fractures are unrelated to the later-formed tectonic fractures in the core, including the high-angle, calcite-mineralized extension fracture ("Ca") and a high-angle strike-slip shear fracture ("strike-slip"). None of the fractures are related to the present-day maximum horizontal compressive stress, which is nearly normal to the strike of the shears as documented by the induced petal fracture strike ("P"). Vertical 4-inch diameter core; uphole is towards the top of the left photo, and away from the viewer in the right photo.

Reference

Finley, S.J., and Lorenz, J.C., 1988, Characterization of Natural Fractures in Mesaverde Core from the Multiwell Experiment. Sandia National Laboratories Technical Report SAND88-1800. Available through the US Government Office of Science and Technology Information: www.osti.gov.

C8

Compound/Reactivated Fractures

Fractures commonly display evidence for superimposed events, both geochemical events (addressed in Section D), such as mineralization and dissolution, and mechanical strain events described here. Perhaps the most common type of fracture reactivation is the crack-seal phenomenon where extension fractures have been repetitively opened and mineralized (e.g., Laubach et al., 2004), but shear fractures can also be continually sheared and mineralized (Figures C8.1, C8.2).

Extension fractures may also be reactivated in shear (Figures C8.3, C8.4), and shear fractures may be reactivated in extension. Extension fractures can be subsequently stylolitized (see Figures C11.7–C11.9), and stylolites can be opened in extension (Figure C8.5). Given the range of structural variation that is possible in geology, any number of cycles and combinations is possible.

Successive chemical events or structural deformations can destroy the evidence for previous events, and reactivated fracturing is probably more common than can easily be documented; it is certainly more common than suggested by the few examples offered here. Given the difficulty in documenting fracture reactivation in core, the dearth of examples reported from image logs is not surprising.

Nevertheless, the recognition and proper interpretation of reactivated fractures can offer insights into the history of the structural development of a reservoir, and can provide a better understanding of both individual fracture permeabilities and the effectiveness of fracture-permeability networks.

Figure C8.1 A repeatedly sheared and calcite-mineralized shear plane in a calcite-cemented sandstone, from the forelimb of an asymmetric anticline draped over a basement-cored thrust fault. Later shear commonly followed the original break in the rock but at one point it veered to a slightly different plane. Sheared mineralization is indicated by the varying fracture width and the different number of calcite layers present at different locations along the fracture. Vertical 4-inch diameter core; uphole is towards the top of the photo.

Figure C8.2 Three views of a subtle, bed-parallel shear plane in a muddy carbonate. The top photo shows the minimal expression of the shear plane on the slabbed surface of the core. The lower left photo shows superimposed oblique slickenlines on the glossy shear-fracture surface, representing at least two events of shear with offset in different directions. The lower right is a close-up view of the superimposed slickenline orientations. Vertical, 3-inch diameter core, uphole is towards the top of the upper photo, and towards the viewer in the two lower photos.

Figure C8.3 Three views of two calcite-mineralized, high-angle extension fractures (X and Y) that have been reactivated and connected by an irregular, intermediate-angle strike-slip shear fracture. The inclined shear plane is ornamented by oblique slickenlines (parallel to the double-headed arrow in the lower right photo). See C8.4 for additional examples of sheared extension fractures. Butts from vertical 4-inch diameter core; uphole is towards the top of the upper and lower left photos; uphole is away from the viewer in the lower right photo.

Figure C8.4 Two views of calcite-mineralized, high-angle extension fractures (the black bar intervals in the left photo) that were reactivated and connected by shear planes (red bar intervals). Subtle, oblique slickenlines (parallel to the double-headed blue arrow in the right photo) are the primary fractography in the shear intervals, and have been superimposed onto the irregular plume structure in the extension fracture faces. Plume structure has been obscured by both the slickenlines and patchy white clay mineralization. The two extension fractures are parallel to each other but offset laterally by two centimeters. The right photo is a close-up of the lower half of the left photograph, the scale bar is in inches. Butts from vertical 4-inch diameter core; uphole is towards the top of both photos.

Figure C8.5 A high-angle stylolite (between the arrows), part of which has been reactivated in extension and filled with calcite. Core is cut from a fine-grained limestone. This superposition of structures implies that the high maximum horizontal compressive stress under which the stylolite formed diminished in magnitude to become the minimum horizontal stress, allowing formation of the extension fracture along and parallel to the stylolite. Vertical 4-inch diameter core; uphole is towards the top of the photo.

Reference

Laubach, S.E., Reed, R.E., Olson, J.E., and Bonell, L.M., 2004, Coevolution of crack-seal texture and fracture porosity in sedimentary rocks: cathodoluminescence observations of regional fractures. Journal of Structural Geology, 26, 967–982.

C9

Shattered Rock

Some cores are heavily fractured yet the rock fragments between the fractures remain in their original positions relative to each other despite significant folding and faulting of the strata. Bedding as well as sedimentary structures are still recognizable since the rock fragments did not rotate or degenerate into grain flow during deformation.

Locally this shattering is due to a process of sequential fracturing and cementation of the individual fractures such that the rock never consisted of disaggregated grains (Figures C9.1, C9.2). Elsewhere the fragments were confined by the in situ stresses so that displacement and rotation of the fragments relative to each other could not occur (Figure C9.3).

Figure C9.1 Core from a limestone shows numerous narrow fractures that probably formed sequentially and were quickly mineralized, healing the rock. The core is oriented, so true south is known. The view is of the bottom of the core looking uphole so that what appear to be WSW-ENE striking fractures are actually ESE-WNW fractures ("120°"). Vertical, 4-inch diameter core; uphole is away from the viewer.

Figure C9.2 Multiple narrow extension fractures formed in this fine-grained, shaley limestone yet bedding is still recognizable and the fracture-bound blocks are not displaced relative to each other. Mineralization healed the rock quickly after each fracture formed. Slab from a vertical, 4-inch diameter core; uphole is towards the top of the photo.

Atlas of Natural and Induced Fractures in Core, First Edition. John C. Lorenz and Scott P. Cooper.

Figure C9.3 Numerous unmineralized parallel fractures in a core cut from vuggy limestone in a heavily folded and fractured reservoir. Fracturing cut the rock into numerous pieces but the pieces were held together by the in situ compressive stresses and the rock did not become disaggregated. Slab surface of a 4-inch diameter core; uphole is towards the top of the photo.

C10

Karst Breccias

Several types of breccia form in karsted terrains. Loucks (1999) has classified them as "crackle," "mosaic," and "chaotic" breccias based on studies of karst/cave systems in the Ellenberger Formation of West Texas. Loucks' first two breccia types form in cave roofs and walls where the unsupported rock begins to fracture and fail due to gravitational pull, but where the rock fragments are still in place relative to each other (Figure C10.1). When the fragments fall to the cave floor, they form chaotic breccias of displaced fragments that can no longer be fit back together (Figure C10.2), and where they may become reworked and rounded. Breccias can become cemented together, forming a rock that can in turn be fractured anew.

The dissolution associated with karst can mix breccias with the clay residues, including residues left by stylolitization. Stylolites may be marked by striations along the teeth created by the movement of the rock past itself during dissolution (Figure C10.3) and which should not be mistaken for fracture slickenlines.

Various types of breccia, including sedimentary breccias, fissure-filling breccias, fault-related breccias, and karst breccias, are not always easy to distinguish from each other given the limited characterization possible with one-dimensional, 4-inch diameter core. Recognition of karst breccias depends in part on evidence for dissolution and the distinctive characteristics of in-place, fracture-bound crackle breccia rock fragments, and in part on context, i.e., whether or not the strata are susceptible to dissolution and whether they are known or likely to contain caves and karst.

Atlas of Natural and Induced Fractures in Core, First Edition. John C. Lorenz and Scott P. Cooper.

Figure C10.1 Mosaic breccias from the roofs and walls of caves formed in a karsted limestone. When cemented together, such breccias form rock that is itself capable of fracturing anew when strained. Vertical, 2-inch diameter cores; uphole is towards the top of both photos (the inverted numeral 3 notwithstanding).

Figure C10.2 Chaotic, cave-bottom karst breccias formed in a chalk, with (left) and without (right) significant postbrecciation cement. Some of these breccias resemble, and may in fact be related to, fissure fills which would also be expected in karst terrains. Left: quarter-section of 4-inch diameter core; right: slab of 3-inch diameter core; uphole is towards the top of both photos.

Figure C10.3 Insoluble black clay and organic remnants from stylolites and dissolution seams in a karsted limestone. Dissolution was associated with shear offset and slickenline formation along vertical stylolite teeth as the rock dissolved and collapsed into itself. Vertical, 2-inch diameter core; uphole is towards the top of both photos.

Reference

Loucks, R.G., 1999, Paleocave carbonate reservoirs: origins, burial-depth modifications, spatial complexity, and reservoir implications. AAPG Bulletin, 83, 1795–1834.

C11

Pocket-Size Geomechanical Systems

Some structural systems are composed of several interrelated elements that are small enough to be captured by a core. These systems can be useful in reconstructing the structural development of a reservoir, and some can be pervasive enough to be pertinent to reservoir permeability. It is also fun just to decipher the composite geomechanical puzzles.

One small-scale, multi-component geomechanical system consists of (1) very short, strata-bound extension fractures that were reactivated and widened by (2) bed-parallel shear, to form (3) millimeter-scale, trapezoidal pull-aparts in a thinly bedded, folded, siltstone-shale. The system was cored on the flank of an anticline (Figures C11.1–C11.3).

The trapezoidal voids opened by this system are lined with calcite, and calcite mineralization is also present locally along the bed-parallel shear planes. The calcite-lined voids were later filled with bitumen, indicating that they are interconnected. The cored strata are slightly inclined due to the anticlinal folding, and the sense of motion recorded by slickenlines on the sheared bedding planes is consistent with flexural slip during folding, i.e., the top blocks moved up-dip.

Similar trapezoidal voids, the sides formed by two high-angle shears and the top and bottom formed by extensional parting along bedding were presented in section B3g. Trapezoidal voids can also form in the horizontal plane, bounded laterally in one direction by bed-normal shear planes and in the other direction by the faces of widened bed-normal extension fractures.

Another pocket-size geomechanical system (Figures C11.4–C11.6) is almost the inverse of the previous example, consisting of:

- a bed-normal stylolite that terminates at a bedding plane
- shear along the bedding plane, but only on one side of the stylolite
- an induced petal fracture that strikes normal to the stylolite.

The petal fracture records a present-day maximum horizontal in situ compressive stress that has the same orientation as the stress that would have formed the vertical stylolite, suggesting that stress system has not changed significantly since stylolite formation. In this example, the sheared bedding plane displays slickenlines that trend normal to the strike of the vertical stylolite, and parallel to the strike of the petal fracture. The bedding plane is sheared on one side of the stylolite but not on the other. Shear accommodated the volume loss along the plane of the stylolite.

The final geomechanical system presented here (Figures C11.7–C11.9) is similar to the previous one but with a complication. The system initially consisted of short, strata-bound, probably mineralized, high-angle extension fractures in a thinly bedded marine shale. However, an exchange of the in situ maximum and minimum horizontal compressive stresses removed the mineralization and most of the original plume structure by pressure solution, forming stylolites along what had been extension fracture planes. Remnants of the original plume structure mark some of the stylolite surfaces.

Small bed-parallel shears, with slickenlines oriented normal to the stylolitized fractures, formed to accommodate the fracture-parallel volume loss during stylolitization. Induced petal fractures confirm that the present-day maximum horizontal compressive stress is normal to the stylolitized extension fractures.

Atlas of Natural and Induced Fractures in Core, First Edition. John C. Lorenz and Scott P. Cooper.

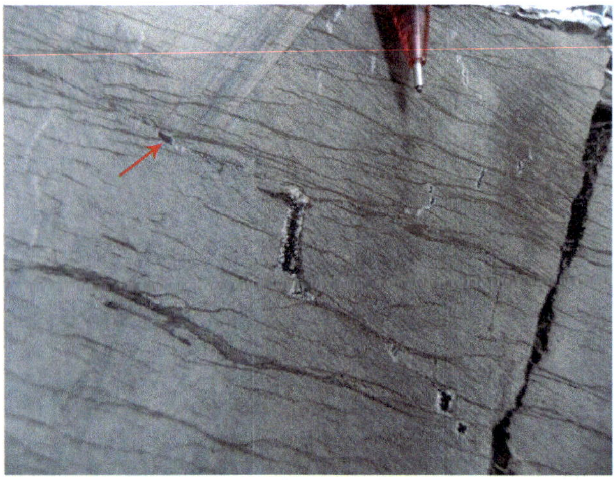

Figure C11.1 Numerous small, white, vertical, calcite-mineralized extension fractures scattered within a thinly laminated siltstone-shale sequence. Surfaces of the bed-parallel shears are not exposed, but shear is inferred from the bed-parallel calcite mineralization present along bedding adjacent to the trapezoids, and from the trapezoidal shape of the voids created by widening of the extension fractures. Arrow points to the trapezoid shown in close-up in the next photo. Slabs from a vertical, 4-inch diameter core; uphole is towards the top of the photo.

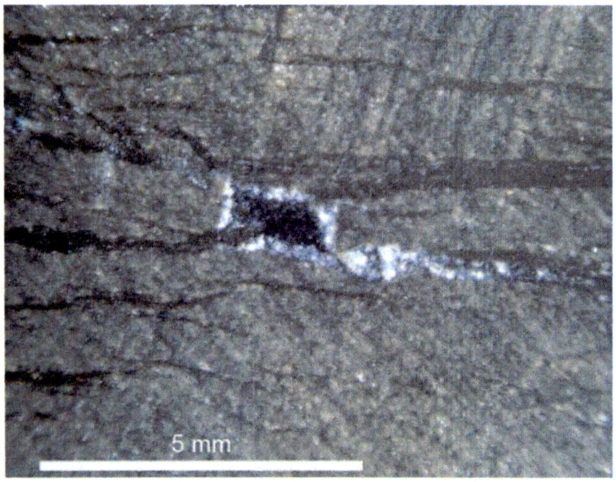

Figure C11.2 Close-up photo of one of the trapezoidal voids formed by widening of an extension fracture in the course of folding the strata during flexural shear slip along bedding planes. The apparent width of the trapezoid in this photo is about 50% more than actual width due to the oblique cut of the slab plane across the void. The opposing faces of a high-angle extension fracture were moved apart during flexural slip to form the lateral walls of the void, which is elongate and forms an irregular, square-sided tube in the third dimension. The top and bottom of the void are formed by the bed-parallel shear surfaces.

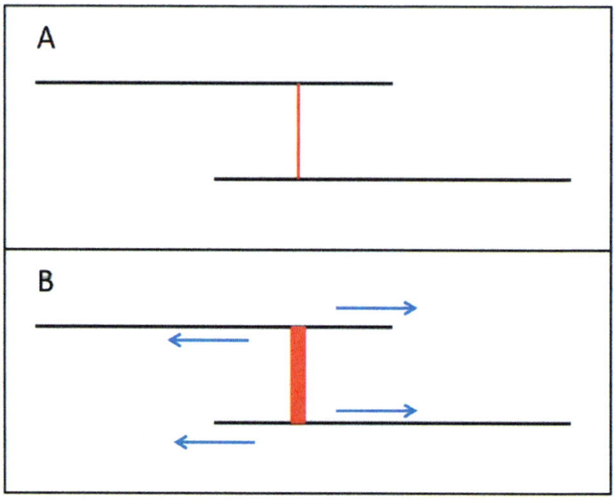

Figure C11.3 Sketch explaining figures C11.1 and C11.2, showing the original bed-bounded high-angle extension fracture (A) and the widening of the fracture by shear along the bounding bedding planes (B).

Figure C11.4 A vertical stylolite (between the arrows) that terminates downward at a sheared bedding plane just above the depth marking. See the following photo for a view of the bedding plane. Vertical 4-inch diameter core; uphole is towards the top of the photo.

Figure C11.5 The section of the bedding plane to the left of the stylolite is subtly slickenlined, parallel to the red arrow drawn on the shear surface, with slickenlines trending normal to the strike of the stylolite. Bedding-plane shear accommodated the loss of material along the stylolite. The bedding plane is not sheared to the right of the stylolite. Vertical 4-inch diameter core; uphole is towards the top of the photo.

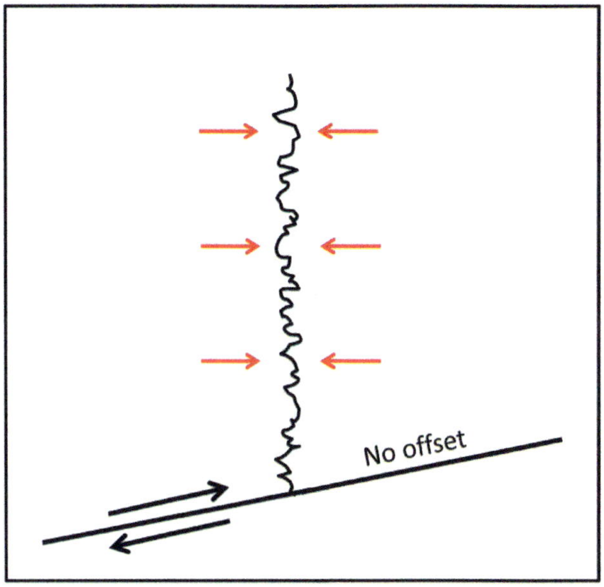

Figure C11.6 Sketch of the geomechanical system shown in Figures C11.4 and C11.5, consisting of an incompletely sheared bedding plane and a vertical stylolite. Shear allowed the block above the bedding plane and left of the stylolite to move to the right far enough to accommodate volume loss during stylolitization. Offset was probably of the order of a millimeter.

Figure C11.7 Short, strata-bound, high-angle fractures viewed edgewise (left), parallel to the white lines, and face-on (right) in core from a thinly bedded, siliceous marine shale. The two fracture surfaces on the right display (1) the remnants of plume structure in the lower fracture, and (2) a poorly developed stylolite surface, that destroyed the plume, on the upper fracture face. Vertical, 3-inch diameter core; uphole is towards the top of both photos.

Figure C11.8 Horizontal and inclined, slickenlined surfaces connect the tips of the stylolitized vertical-extension fractures ("SVE"), accommodating volume loss along the stylolites. An unrelated, calcite-mineralized vertical extension fracture ("VE") is also present in the core. Vertical, 3-inch diameter core; uphole is towards the top of the lower photo, and towards the viewer in the upper photo.

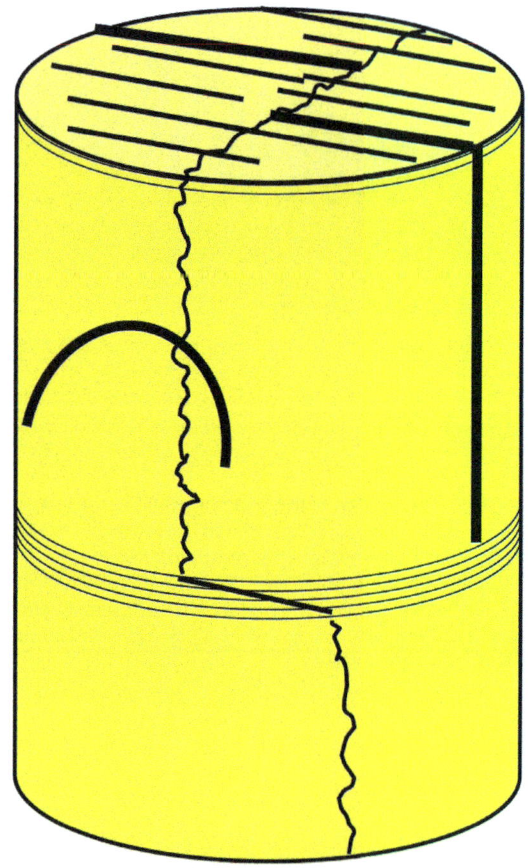

Figure C11.9 A geomechanical system composed of stylolitized vertical extension fractures (irregular lines), later vertical extension fractures (heavy black lines), and horizontal and inclined shear along slickenlined planes along and oblique to bedding, accommodated the volume loss along the stylolites explaining Figures C11.7 and C11.8. A petal fracture (curved black line) records a present-day maximum horizontal compressive stress in the direction that would have created pressure solution and stylolitization along the previous extension fracture face.

C12

Stylolites

Stylolites, formed by dissolution under asymmetric confining stresses and lined with insoluble materials, can be considered to be fractures in the sense that they are irregularly planar features formed during deformation of the rock. Moreover, although they rarely display obvious apertures, Nelson (1981) and Wennberg et al. (2016) have both suggested that stylolites can provide permeability pathways in a reservoir. In addition, the extension fractures that form along some stylolites are important contributors to a reservoir permeability system. Where they are well developed, these fractures extend for several centimeters or even a few tens of centimeters normal to a stylolite plane, forming systems of short fractures commonly with subparallel, irregular strikes adjacent to the stylolite. Each stylolite that has associated, well-developed fractures can create a permeability streak in a reservoir, and stacked stylolite-fracture systems can contribute significantly to reservoir plumbing.

It is not always easy to determine whether a set of extension fractures originated at a stylolite or whether fractures predate a stylolite and were truncated by the associated dissolution. However, there are some unique characteristics that help with this differentiation. Stylolite-related extension fractures are typically widest at the stylolite and commonly originate at stylolite teeth. The fractures taper to blind terminations within the adjacent rock. They may be mineralized but they are commonly incompletely filled. They may be developed on one or both sides of a stylolite. Stylolites and the related fractures are most common in relatively soluble carbonate strata, but they also occur in sandstones, particularly if the sandstones have been deeply buried or subjected to significant tectonic compression.

Figure C12.1 Two views of a well-developed, bed-parallel stylolite in a muddy limestone, with a significant thickness of black insoluble material concentrated along the stylolite seam but no associated fractures. A quarter section of a vertical, 4-inch diameter core; uphole is towards the top of the left photo and away from the viewer in the right photo.

Atlas of Natural and Induced Fractures in Core, First Edition. John C. Lorenz and Scott P. Cooper.

Figure C12.2 The characteristic pitted and embayed surface shown here illustrates a bed-parallel stylolite in limestone. Without the characteristic remnant insoluble residues, a stylolite may be easily missed when logging a core. Residues can be lost during core handling or they may adhere to only one stylolite face which has been removed for sampling. Butt from a vertical 4-inch diameter core; uphole is away from the viewer.

Figure C12.3 Some stylolites are weakly developed, as shown by this dimpled surface of an incipient stylolite in a limestone. It is barely noticeable in cross-section on the slab face yet it is one of several within a 1-foot section of core. Vertical 4-inch diameter core; uphole is away from the viewer.

Figure C12.4 Well-developed stylolites are easily recognizable even in core rubble by their tall teeth. They often cause a core to disintegrate during core processing, leaving only the slickenlined surfaces of the teeth as evidence for stylolitization. Vertical 4-inch diameter core; uphole is towards the top of the left photo, and is unknown, either toward the top or the bottom, of the right photo.

Figure C12.5 The most common orientation for stylolites is horizontal and parallel to bedding, formed where the weight of the overburden provided the maximum compressive stress. Vertical, bed-normal stylolites can also form, typically in structural settings such as thrust belts where the maximum compressive stress lies in the horizontal plane. Stylolites may be tilted to inclined positions when bedding is folded. Superimposed stylolites such as those in this example in core from a chalk record multiple stress conditions. Quarter section of a vertical, 4-inch diameter core; uphole is towards the top of the photo.

Figure C12.6 In thinly laminated carbonates, high-angle stylolite surfaces may form as bed-parallel ridges rather than conical teeth. Vertical, 4-inch diameter core; uphole is towards the top of the photo.

Figure C12.7 Stylolites are most common in carbonates but they can also form in deeply buried sandstones. This quartz sandstone core containing low-amplitude, irregularly horizontal stylolites was cut from a depth of 6463 ft but was at one time buried to over 14 000 ft. Kaolin-filled fractures developed along the stylolites, commonly emanating from the teeth and extending a few centimeters into the adjacent rock before wedging out and terminating. Thin sections show remnant porosity within the kaolinite filling. Vertical 3-inch diameter core; uphole is towards the top of the photo.

Figure C12.8 A network of interconnected, stylolite-related fractures adjacent to a stylolite and exposed on horizontal and vertical planes, accentuated by pencil lines drawn adjacent to each fracture. Some stylolite-related fracture systems are formed of more parallel fractures. Vertical, 4-inch diameter core; uphole is towards the top of the photo.

Figure C12.9 A set of roughly parallel stylolite-related fractures ("sty gashes") associated with a horizontal stylolite in a limestone. The fractures extend only a few centimeters up and down from the stylolite before terminating, but they are well developed, interconnected, and poorly mineralized, and they should form a horizontal permeability streak in the reservoir. Vertical, 4-inch diameter core; uphole is towards the viewer.

Figure C12.10 Much of the material dissolved from a rock to create a stylolite is precipitated in the pore space adjacent to the stylolite, creating low-porosity zones along stylolites as illustrated by the exclusion of the brown oil stain in this core from the calcarenite rock near the stylolite. The distribution of oil staining also reflects higher porosity along the stylolite-related fractures and perhaps along the stylolite itself. Slab of a 4-inch diameter core; uphole is towards the top of the photograph.

Figure C12.11 The higher porosity of stylolite-related extension fractures is illustrated by these two photos of a limestone core. The fractures exposed on the slab face (left) contain oil, as shown by fluorescence when exposed to ultraviolet light (right; preslabbing, whole core in the same interval). Vertical, 4-inch diameter core; uphole is towards the top of both photos.

References

Nelson, R.A., 1981, Significance of fracture sets associated with stylolite zones. AAPG Bulletin, 65, 2417–2425.

Wennberg, O.P., Casini, G., Jonoud, S., and Peacock, D.C.P., 2016, The characteristics of open fractures in carbonate reservoirs and their impact on fluid flow: a discussion. Petroleum Geoscience, 22, 91–104.

Section D

Mineralization

D1

Mineralization

D1a Introduction

As noted by Wennberg et al. (2016), fracture models based on the varying mechanical properties of different layers in a stratigraphic sequence may produce reasonable estimates of fracture-intensity distributions in a reservoir, but they do not account for the postfracture mineralization and/or dissolution that strongly affect the permeability of individual fractures. Mineralization restricts fracture-parallel permeability by narrowing apertures while dissolution associated with diagenesis can widen fracture apertures and enhance permeability. Many fractures exhibit evidence for both processes. Mineralization may or may not also restrict permeability across a fracture face, inhibiting flow from the matrix into a fracture, depending on the ratio of matrix permeability to that of the layer of mineralization. Lorenz et al. (1989, 2005) measured the permeability of narrow calcite-occluded fractures in the laboratory and found that permeability across and along the fractures equaled or exceeded the microdarcy-scale matrix permeability of the tight, natural gas sandstone reservoirs, so some types of mineralization at least do not restrict fluid flow in some types of reservoirs.

The importance of a fracture system to reservoir plumbing therefore depends on the ratio of fracture permeability to matrix permeability. A fracture system with microdarcy-scale fracture-parallel flow would be insignificant in a reservoir with millidarcy matrix permeability, but the same microdarcy fracture system would be an important production mechanism in a resource-play reservoir with nanodarcy-scale matrix permeability, especially if reservoir pressures are high and the hydrocarbon being produced is gas rather than oil.

Unmineralized natural fractures are common, but most natural fractures have been altered by at least one mineralization event, and many display superimposed mineral layers. Dissolution may be more common than is generally recognized since minor dissolution on a fracture face can be difficult to distinguish from faces that have never been mineralized, and dissolution is often obscured by a layer of later mineralization. Dissolution can also remove earlier layers of mineralization.

Minerals may be precipitated in a fracture at any time during or after fracturing depending on the burial and geochemical histories of a formation. Fracture diagenesis may even occur during the short life span of an oilfield, with mineralization precipitating in a fracture due to the changing pressures and temperatures associated with production, in the same way mineral scale is deposited in production tubing. An internet search of "wellbore tubing scale" brings up impressive images of the flow-restricting mineralization inside the pipes of oilfield collection systems.

Data on fracture widths and apertures in many cores suggest that fractures with larger widths are commonly more completely mineralized than narrower fractures in terms of percent occlusion (Figure D1a.1), but a small

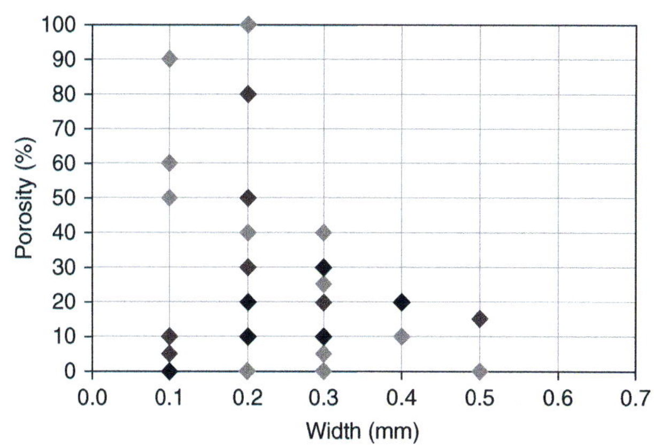

Figure D1a.1 Cross-plot comparing fracture width to percent remnant fracture porosity within the mineralized fracture width, for high-angle extension fractures in two calcareous-shale formations. The data show a trend that is common in many fracture data sets, i.e., that wider fractures tend to have a larger percentage of their widths occluded by mineralization. However, even a small percentage of remnant porosity within a wide fracture can be an effective conduit to fluid flow. The data also show that narrow fractures in many formations can retain significant remnant porosity and can contribute to a reservoir plumbing system (n = 53). Darker points highlight overlapping data.

Atlas of Natural and Induced Fractures in Core, First Edition. John C. Lorenz and Scott P. Cooper.

percentage of remnant fracture porosity in a wide fracture can be more important than a larger percentage remnant porosity in a very narrow fracture. Wider fractures are commonly taller and longer, and until mineralized they would have been more accommodating to fluid flow than smaller fractures. Small fractures are still important, however, since there are typically many more small fractures than large fractures in a system. Moreover, whole-core permeability measurements commonly indicate that even narrow, small fractures can retain a significant degree of effective permeability.

D1b Calcite Mineralization

Calcite is one of the more soluble and mobile components in sedimentary systems and it may be the most common type of fracture mineralization in hydrocarbon reservoirs. Other minerals that are commonly found in fractures include, in no particular order, quartz, clay, pyrite, barite, dolomite, bitumen, gypsum, and halite.

Calcite mineralization commonly displays a crystalline habit with euhedral faces (Figure D1b.1), evidence that the mineralization grew into open void space within the fracture width. Calcite occurs both as continuous layers covering fracture faces and as scattered crystals (Figures D1b.2, D1b.3). Mineralization in larger fractures typically grew from the walls inward towards the middle of the aperture (Figure D1b.4), but examples of single-layer mineralization are also common.

Calcite mineralization can occur as macroscopically amorphous layers (Figure D1b.5), particularly where it grew into fracture apertures too narrow to allow development of well-defined crystals (Figures D1b.6, D1b.7). Amorphous calcite may also be the result of postprecipitation shear (Figure D1b.8). If shear was minimal, the calcite may form a fish-scale pattern that records the direction of shear (Figure D1b.9).

Acicular, prismatic crystals are usually interpreted to have grown concurrently with fracture opening (Figure D1b.10). Some mineralization layers appear to have been recrystallized after precipitation so that large patches of the mineral are in crystallographic continuity along the fracture face (Figure D1b.11). All of these textures affect the permeability of a fracture and should be noted when logging.

Figure D1b.1 Euhedral crystalline calcite covers the face of an extension fracture in core cut from an overpressured fluvial sandstone at 6200 ft depth. In 1960, Griggs and Handin, respected experts from the Shell rock mechanics laboratory, wrote that "It is inconceivable that an open crack can be present at depth…", on the assumption that the high confining stresses at depth would quickly close any planar voids. That was written prior to a fuller understanding of the concepts of effective stress (i.e., Gretener, 1979), and the role that pore pressure plays in counteracting a large percentage of the in situ confining stresses. Crystalline mineralization such as that illustrated here is common in fractures, and had to have grown into an open fracture aperture. The bridge of rock in the center of both photos, which originally connected the opposing fracture faces, is surrounded by white rather than translucent calcite, suggesting that it has been strained, possibly by bending the web of rock during fracture reactivation after mineralization. Vertical 4-inch diameter core; uphole is towards the top of the upper photo, and towards the viewer in the lower photo.

Mineralization can be irregularly distributed along a fracture such that it is difficult to estimate the degree of remnant fracture porosity for the entire cored fracture (Figure D1b.12), but care must be exercised since core processing can affect the apparent degree of mineralization in a fracture (Figures D1b.13, D1b.14).

Figure D1b.2 Some mineralization consists of isolated calcite crystals such as those scattered on this high-angle, extension-fracture face. Although they obstruct very little of the fracture width in this sandstone, they would create turbulence in the flow of fluid along the fracture plane, and they provide an indication of the minimum fracture aperture. Vertical, 4-inch diameter core; uphole is towards the top of the photo.

Figure D1b.3 Mineralization can vary along a fracture face as a function of lithology. This example shows subtle variations in calcite mineralization along the face of a high-angle extension fracture in a marine shale, with small calcite nodules dotting the face of the fracture where it cut across the two thin calcareous layers (white brackets). The rest of the fracture, which cuts less-calcareous shales, is unmineralized. Vertical, 4-inch diameter core; uphole is towards the top of the photo.

Figure D1b.4 Three fracture-related faces (parallel to the plane of the photograph) in a marine shale were formed by the growth of mineralization from opposing walls into a millimeter-scale open fracture aperture at depth. The white arrow shows the fracture face, from which mineralization has spalled off and is missing; the red arrow shows the euhedral crystalline surface of the calcite that grew outward from that fracture face towards the viewer; the black arrow indicates the back of the second, opposing mineralization layer, which had been attached weakly to the missing fracture face and which grew away from the viewer into the fracture aperture. Vertical, 4-inch diameter core; uphole is towards the top of the photo.

Figure D1b.5 Calcite mineralization is not always white; the calcite covering this fracture face in a marine shale is oil-stained yet effervescent. Half of the slab of a vertical, 4-inch diameter core; uphole is towards the top of the photo.

Figure D1b.6 Some extension fractures have widths of less than a tenth of a millimeter and the width is relatively uniform along the height and length of the fracture. This narrow extension fracture in a marine shale was filled with calcite that was weaker than the rock, so the core broke along the fracture plane. The thin calcite layer, flaking off the fracture surface in small white patches, offers no indication that the mineralization grew inward from the opposing walls. Twist hackle at the top of the fracture and the plume structure on the fracture face both can be discerned underneath the thin layer of calcite. Vertical, 4-inch diameter core; uphole is towards the top of the photo.

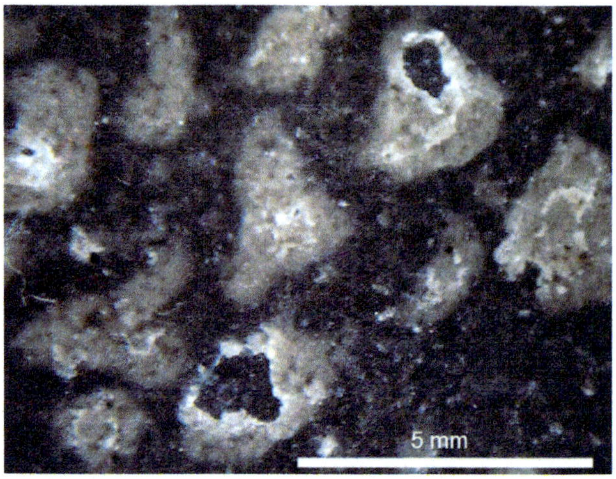

Figure D1b.7 Some thin layers of calcite have a bubbled texture as on the face of this extension fracture (parallel to the plane of the photographs) in a marine shale. The texture almost suggests that natural gas became trapped beneath the mineralization as it diffused out of the core, ballooning the layer of calcite. Vertical, 4-inch diameter core; uphole is towards the top of both photos.

Figure D1b.8 Two views of calcite mineralization in a shear fracture. Left: edge-on. Right: the fracture face with slickenlines underlying the calcite mineralization. Shear fractures commonly have large voids into which euhedral mineralization can be precipitated. The largely amorphous nature of the calcite in this fracture suggests that it was sheared after precipitation. Vertical, 4-inch diameter core; uphole is towards the top of both photos.

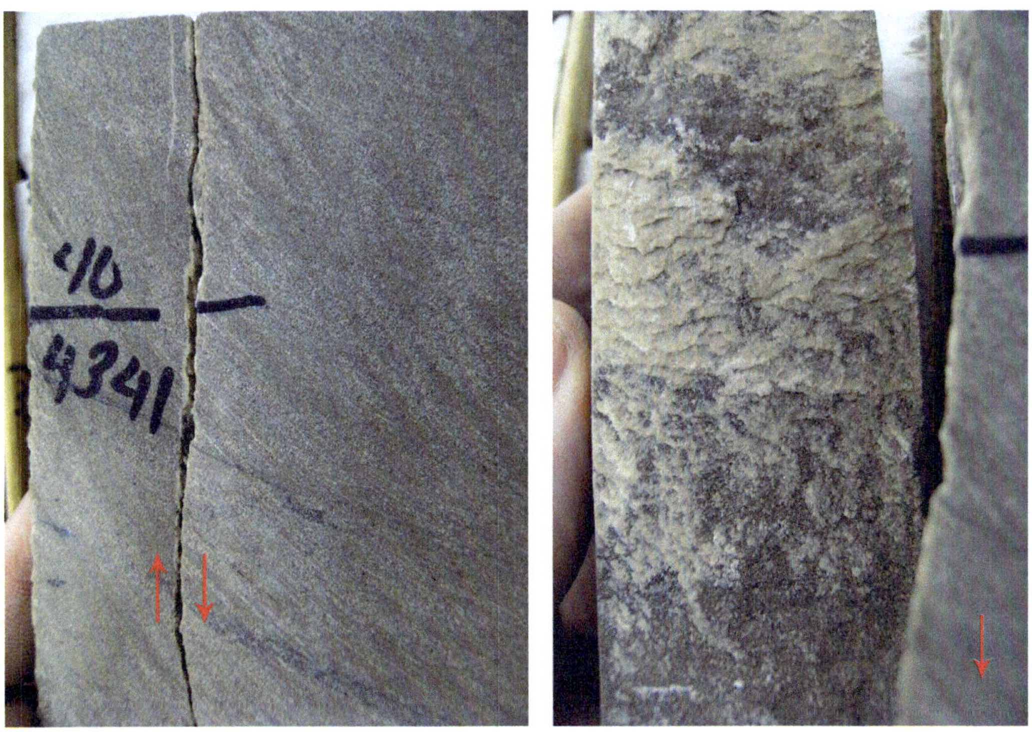

Figure D1b.9 Shear offset of the order of a millimeter can produce a fish-scale texture in calcite fracture mineralization. These two photos show the edge-on view (left) and face (right) of a high-angle fracture mineralized with fish-scale calcite. Bedding offsets indicate that the asymmetry of mineralization records the direction of offset, as indicated by the red arrows. Slabs of vertical, 4-inch diameter core; uphole is towards the top of both photos.

Figure D1b.10 Parallel, acicular crystals are most common within horizontal "beef"-filled extension fractures, suggesting a unique origin for these structures. The acicular habit is usually interpreted to indicate crystal growth concurrent with fracture opening, with no open void space that would have allowed for the development of euhedral crystal faces. Crystals can grow displacively within unconsolidated sediment, but fractures in lithified rock are not pushed open or held open by the pressure of crystallization since point-contact pressure increases the solubility of crystals and the pressure-solution effects exceed the force of crystallization. Vertical, 2.5-inch diameter core; uphole is towards the top of both photos.

Figure D1b.11 Some single-layer fracture mineralizations have been recrystallized, filling the fracture width with relatively large patches of single-crystal but non-euhedral calcite. The extension fracture shown here (parallel to the plane of the photo) formed in a dolomite and is covered by a crystallographically continuous calcite layer. Broken edges of the layer form geometric, right-angle edges along the crystallographic axes. The red-black line pair indicates that uphole is towards the top of the photo; the green line is a master orientation line, useful for recording the continuity of core orientation prior to sampling and slabbing. Vertical, 4-inch diameter core; uphole is towards the top of the photo.

Figure D1b.12 Mineralization is not always uniformly distributed along a fracture. This photo shows the opposing faces of an extension fracture in sandstone. Patchy white calcite mineralization is better developed near the bottom of the fracture where its width narrows toward its termination against a shaley layer (not shown); mineralization is poorly developed higher in the fracture. The calcite patches do not correlate across the fracture, suggesting that they grew inward from the opposing faces. Vertical 4-inch diameter core; uphole is towards the top of the photo.

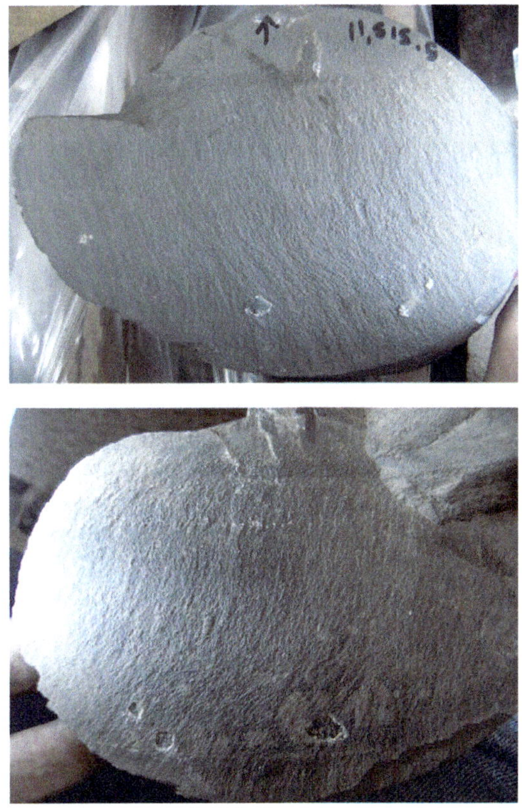

Figure D1b.14 In extreme examples, such as the two faces of the high-angle extension fracture shown here, cut from a fine-grained limestone, almost all of the thin layer of calcite mineralization in a narrow fracture adheres to one face when the rock breaks along the fracture plane. The upper photo shows several small patches of calcite on the otherwise unmineralized fracture face; the lower photo shows the opposing face with a nearly complete calcite layer. Horizontal, 4-inch diameter core with the fracture cutting oblique to the core axis. Stratigraphic up (indicated by the black arrow) is towards the top of both photos.

Figure D1b.13 Care must be taken in assessing the degree of mineralization in a fracture aperture. This fracture plane in a well-cemented sandstone has been opened and is displayed in butterfly fashion with the opposing faces parallel to the plane of the photo and facing the viewer. The white calcite mineralization covers about half of each face, but calcite completely filled the fracture and has been pulled off in patches, some of which stuck to one face, some to the other. Holes in the mineralization on one fracture face match the shape of mineralization patches on the other. If only one of these fracture faces could be examined, the fracture might be mistakenly assessed as 50% occluded. Vertical, 4-inch diameter core; uphole is towards the top of the photo.

D1c Other Types of Mineralization

Although just about any type of mineral can be found filling a fracture, depending on the geochemical history of the formation, after calcite some of the other common types of mineralization are quartz (Figures D1c.1, D1c.2), dolomite (Figures D1c.3, D1c.4), anhydrite (Figure D1c.5), and pyrite (Figure D1c.6).

Many fractures show evidence for multiple mineralization events, which may be more common than recognized since younger mineralization can cover and obscure older layers. Two or more precipitation events may deposit layers of the same basic mineral which can be differentiated by color or crystal habit (Figure D1c.7), while other superimposed layers consist of entirely different minerals (Figures D1c.8, D1c.9). Thin layers of metamorphosed rock commonly form along shear fractures in shales (Figure D1c.10), and although they do not comprise a precipitated mineralization, the layers retard flow from the matrix rock into a fracture aperture.

In sandstone reservoirs in the Rocky Mountain Region of the US, an early druze of submillimeter quartz crystals commonly underlies and is obscured by a later layer of calcite (Figure D1c.11). The crystals of quartz may be too small to be recognized without thin sections and a microscope. However, making a thin section of a fracture requires care; the best angle to view a fracture in thin section is usually in a plane normal to the fracture face, and if the fracture is broken open the fracture face will form one edge of the thin section chip. The important details of the fracture face can be lost at various stages of thin section preparation if this edge is not carefully preserved. Thin section orientations relative to a fracture plane are important and should be recorded as the thin section is made.

Figure D1c.1 Amorphous quartz fills a wide extension fracture with secondary shear offset in a well-cemented siltstone. The top core piece is separated from the rest of the core by a spinoff and cannot be locked in place. The original fracture fill may have been crystalline quartz, altered to amorphous quartz by the low-grade metamorphism evident in this Precambrian quartzitic siltstone. Vertical, 4-inch diameter core; uphole is towards the top of the photo.

Figure D1c.2 The rough face of this natural extension fracture appears to be unmineralized but it is coated with a fine druze of submillimeter euhedral quartz crystals that cannot be seen without a microscope. The facets of the crystals sparkle, reflecting light and giving a clue to their presence, but care must be taken in the interpretation, comparing this surface to a known fresh break in the rock, since broken quartz grains can give similar reflections. Vertical, 4-inch diameter core; uphole is towards the top of the photo.

Figure D1c.3 Large, diamond-shaped crystals of dolomite partially cover the face of this extension fracture in a dolomite. The red-black line pair indicates the uphole direction of the core, the green line is an orientation line. Vertical 4-inch diameter core; uphole is towards the top of the photo.

Figure D1c.4 Dolomite crystals bridge the width of these irregular fractures in a dolomite host rock, producing irregular pathways for fluid flow along the fracture planes. Vertical, 4-inch diameter core; uphole is towards the top of the photo.

Figure D1c.5 Anhydrite was precipitated on this fracture face in an anhydrite-cemented non-marine siltstone. Euhedral crystals formed in zones of wider fracture aperture whereas the anhydrite is amorphous in the narrow parts of the fracture. Vertical, 4-inch diameter core; uphole is towards the top of the photo.

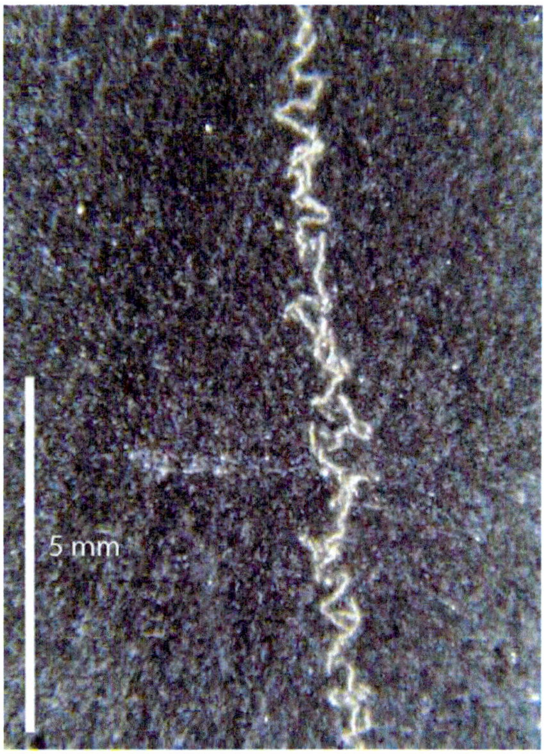

Figure D1c.6 Pyrite is common in fractures associated with hydrothermal fluids and in fractures that formed early during anaerobic diagenesis. This early-formed, ptygmatically folded fracture is filled with pyrite. Vertical, 2.5-inch diameter core; uphole is towards the top of the photo.

Figure D1c.7 Two calcite-mineralization events in a chalk formation are indicated by the difference between the white, crystalline calcite mineralization and the yellow, more amorphous calcite mineralization. The white calcite fill retains significant remnant fracture porosity. The relative ages of the two fracture-filling events are unclear in the left photo, but in the right photo the white calcite fracture filling can be inferred to predate the yellow calcite filling since the white calcite has been truncated by stylolitization but the yellow calcite has not. Quarter sections of vertical, 4-inch diameter core; uphole is towards the top of both photos.

Figure D1c.8 Fractures formed at different times may contain different types of mineralization, as shown in this core from a marine shale. The early, less planar fracture on the right is filled with amorphous gray quartz; the parallel but later, more planar fracture on the left is filled with calcite. Slab from a vertical, 4-inch diameter core; uphole is towards the viewer.

Figure D1c.9 Pyrite can form as patches along early fractures, with later mineralization filling the open fracture aperture around the pyrite. This inclined fracture in marine shale is filled with early pyrite and later calcite. A little pyrite can dominate a fracture signature on an image log. Vertical, 2.5-inch diameter core; uphole is oblique out of and towards the top of the photo.

Figure D1c.10 Slickenlined, glassy shear surfaces in shales may or may not be mineralized, but the alignment of clay particles produced by shear creates an effective mineralization that can retard fluid flow from the matrix rock into the fracture aperture. The broken rock at the lower left of this photo illustrates the limited thickness of the polished layer where it contrasts with the underlying poorly bedded silty marine shale. Vertical 4-inch diameter core; uphole is towards the viewer.

Figure D1c.11 Thin section showing calcite filling the aperture in a cored sandstone, the calcite overlies small quartz crystals that grew into the fracture aperture prior to calcite mineralization. A central zone of hydrocarbon particulates was probably disrupted and displaced into the center of the fracture as mineralization filled the fracture (see Lorenz et al., 1998).

D1d Oil and Bitumen

A wide variety of hydrocarbons can fill fractures, ranging from remnants of the mobile oil that fills a reservoir (Figures D1d.1–D1d.3) to different types of pyrobitumen that are baked oil residues resembling coal (Figures D1d.4–D1d.7). Mobile oil does not occlude fractures and does not degrade permeability, but devolatilized tars and pyrobitumens do and thus can be considered to be a form of mineralization. Consequently, it is important to differentiate the two types of hydrocarbons even though many reservoirs contain both mobile oil and the immobile remnants of earlier oil, making interpretations difficult. A further complication is that oils that were mobile under reservoir temperatures and pressures devolatilize and lose their mobility when brought to the surface and left to sit for years in a core box in a hot warehouse (Figure D1d.8), coming to resemble the less-mobile hydrocarbons that obstruct reservoir permeability.

Some fractures display mixed hydrocarbon and inorganic mineralization (Figure D1d.9). The relative ages of the two phases are not always clear, but the most plausible interpretation is commonly that the hydrocarbon is younger, filling intercrystalline voids as it migrated into the incompletely mineralized fracture. Nevertheless, it is not impossible for inorganic mineralization to develop in fracture apertures, albeit slowly, underneath previously emplaced hydrocarbons.

Live reservoir oil, ranging from yellow to black-brown in color and often fluorescing under UV light, will stain hands when logging a core. A core recovered from a reservoir containing relatively light oil may be swimming in oil when first recovered but that oil can evaporate and/or be washed way during core processing. Don’t mistake oil-based drilling mud for reservoir oil.

Thicker oil remnants in the form of tars and bitumens are darker and sticky. They may still ooze from the rock, commonly along fractures, for months after a core is cut. It can be difficult to determine whether such hydrocarbons are remnants of live oil or constitute a fracture-blocking mineralization. In contrast, pyrobitumens definitively occlude fractures. Pyrobitumens may crack and be blocky and lustrous, resembling coal. A black dust may rub off on fingers and hands while handling cores containing fractures filled with pyrobitumens.

Figure D1d.1 A light, live oil from the reservoir coats the calcite-mineralized surface of this high-angle extension fracture captured by core cut from a calcareous marine shale. The oil has a yellowish cast and water beads up on the surface of the fracture (upper, central part of the photo). If a reservoir oil is light it may be flushed from the core during drilling and processing, and if not it may evaporate over the course of hours or days, leaving little evidence for oil in the core. Vertical 5¼-inch diameter core; uphole is towards the top of the photo.

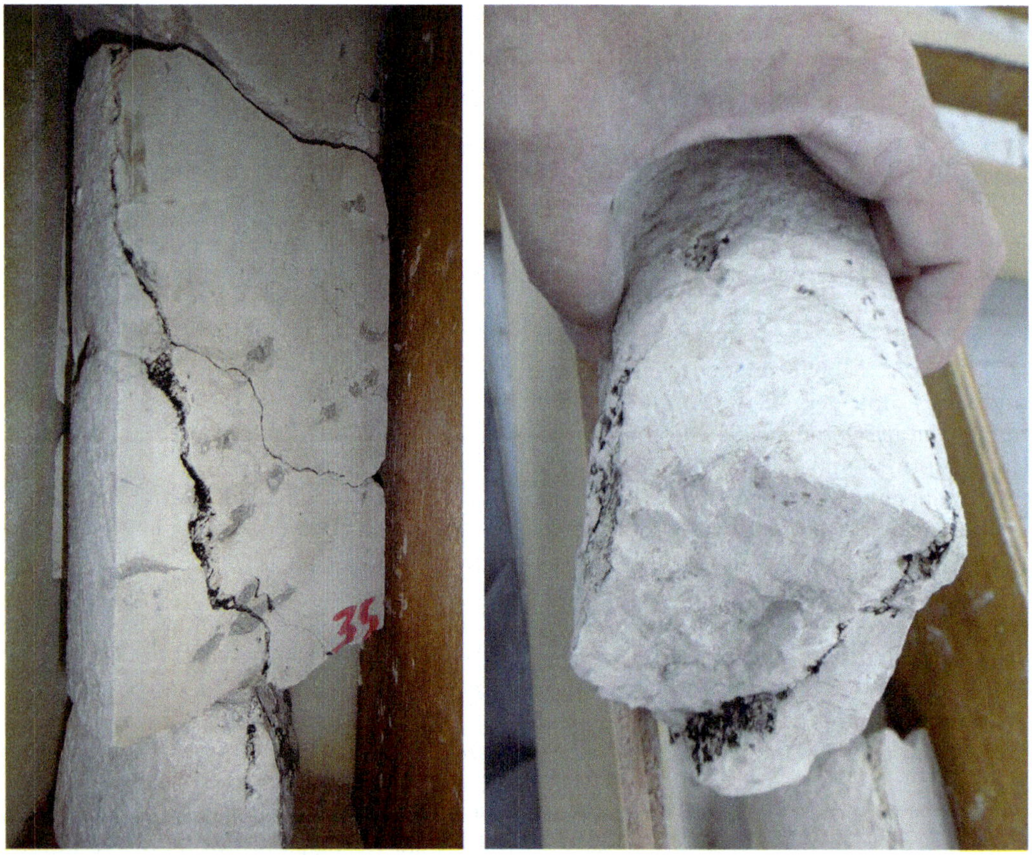

Figure D1d.2 Two views of an extension fracture filled with a thick but live reservoir oil. Live reservoir oil can be distributed pervasively in the rock or may be concentrated in fractures. Thick oils may bleed from a core for months after the core has been cut. The presence of bleeding oil on postcoring surfaces such as slab planes helps distinguish it from immobile bitumen. The thick, sulphurous live oil in this chalky limestone is concentrated in the fractures, and at the time of logging was no longer mobile although it was still sticky. Butts from vertical, 4-inch diameter core; uphole is towards the top of both photos.

Figure D1d.3 The thick oil in this reservoir permeates the eolian sandstone and coats this calcite-mineralized, strike-slip shear fracture surface (parallel to the plane of the photo), obscuring the mineralization, slickenlines, and shear steps. The core had been cut three decades before this photo was taken but the sticky, tarry oil remnants still oozed from the rock and out onto fresh breaks in the core, and still stained fingers a brown color. Fragment of a vertical, 4-inch diameter core; uphole is towards the top of the photo.

Figure D1d.4 This cored fracture in a muddy limestone is lined with a non-sticky, immobile, black, solid bitumen which is the remnant of an early phase of oil that filled the reservoir. A different oil later refilled the reservoir, after this oil had degraded to bitumen, and the later oil is being produced. The early bitumen partially plugs and degrades permeability along the fractures. The drops of muddy water on the fracture face were left after washing drilling mud from the fracture face to expose it. Slabs of vertical, 4-inch diameter core; uphole is towards the top of the photo.

Figure D1d.5 The black, blocky, lustrous hydrocarbon covering this fracture face, parallel to the plane of the photo, is a naturally devolatilized pyrobitumen from a cored lacustrine carbonate sequence. Pyrobitumen commonly resembles coal, with conchoidal breakage that allows the material to crumble into particulates. It is not sticky but does create a black powder that will stain hands. Half of the slab of a 4-inch diameter core; uphole is towards the top of the photo.

Figure D1d.6 Two views of a short, high-angle extension fracture filled with pyrobitumen, cored from a marine limestone. The fracture plane is nearly parallel to the slab face. The lower photo shows both fracture faces, opened up in butterfly fashion with the "hinge" at the bottom of the fracture along bedding. Note the blocky appearance of the broken bitumen (particularly near the arrow). Bedding is inclined, the 4-inch diameter core is vertical, and uphole is towards the top of both photos.

Figure D1d.7 A solid, pyrobitumen partially fills an irregular, high-angle extension fracture in a fine-grained marine limestone. The pyrobitumen-filled fracture intersects a different, more planar fracture that has an unmineralized but aged, brownish surface, parallel to the plane of the photo. Patches of freshly broken gray rock are also exposed. The white line down the axis of the core in the lower part of the photo is a scratch created by a technician who used a knife to cut away the wrapping of plastic film that was used to temporarily preserve fluid saturations. Vertical, 4-inch diameter core; uphole is towards the top of the photo.

Figure D1d.8 It is not always easy to distinguish between pyrobitumen and the remnants of a thick but mobile oil in a core. This 1-inch diameter plug was cut along the axis of a vertical, hydrocarbon-filled extension fracture, and the fracture is opened in butterfly style to show both fracture faces. Although solid and blocky, the hydrocarbon is not lustrous or brittle and is probably not a pyrobitumen. Neither is it a sticky tar or a viscous oil. It is probably the remnants of the high-viscosity oil that fills this reservoir, devolatilized when brought to the surface and during the course of many years sitting in a core box in a hot Texas warehouse. The brownish material at the center of the photo is drilling mud. Horizontal, 1-inch plug, cut from a vertical 4-inch diameter core; the plane of the fracture was vertical in situ, and uphole is either to the left or to the right of this photo.

Figure D1d.9 Two views of a calcite-mineralized and pyrobitumen-filled fracture in a limestone. The calcite and pyrobitumen are inextricably mixed within the fracture width, both of them obstructing the flow of the more mobile reservoir oil. Vertical, 4-inch diameter core; uphole is towards the top of both photos.

D1e False Mineralization

The presence of mineralization would seem to be a definitive criterion for determining that a fracture is natural, but the identification of a substance as mineralization is not always unambiguous. Coring and handling processes create several types of materials and structures that resemble mineralization, and they can be found on the surfaces of both natural and induced fractures.

The easiest to dismiss as false mineralization is the black but coppery pipe dope, the copper-bearing grease used to lubricate joints in a drill string. Blobs of this material invariably get mixed into drilling muds and can be found plastered onto core surfaces and pumped into fracture apertures. Both the dope and sometimes the drilling mud itself can resemble mineralization to an unexperienced eye.

The formation, drilling, or slabbing fluids that permeate a core are commonly saturated with sulfates or chlorides. After a core is cut or slabbed, these fluids evaporate and the minerals they contain can precipitate on any handy core surface, including the faces of natural and induced fractures (Figures D1e.1, D1e.2). Some of these efflorescences form filigrees of fine hair that are obviously not present in situ, but others resemble more solid, natural mineralization. If a fracture surface is coated with a material suspected to be efflorescence, look to see whether that material is also present on obviously artificial surfaces such as slab faces and plug holes.

Some marine fossils, particularly the bivalve *Inoceramus*, are formed of acicular, prismatic calcite crystals oriented normal to the shell surface, and are similar to the crystalline calcite found in horizontal "beef"-filled extension fractures (Figure D1e.3).

Slabbing technicians are usually tasked to slab a core so that it will be photogenic. (This slabbing philosophy leads to preferential relegation of fractures to core butts and problems in determining fracture intensities from slabs alone.) If a slab plane cannot be cut to miss a fracture, technicians have been known to intentionally glue a fracture together so that the core stays intact during slabbing. While this is usually obvious where the glued face is exposed (Figure D1e.4), it can be ambiguous, if not downright misleading, where the glued fracture can only be viewed in cross-section.

When a natural or induced fracture plane is slabbed at a shallow angle, the poorly supported wedge of rock on one side of the fracture is often shattered and cracked for a millimeter or so next to the fracture by either the slab saw or the electric sander used to remove slab-saw scars from a slab surface. The pieces of cracked rock remain in place and can resemble mineralization (Figure D1e.5), but the shattered nature of the structure is easily resolved with a hand lens and by recognizing that the shatter zone is restricted to a depth of less than a millimeter below the slab surface.

Finally, the rock powder created by coring, slabbing, and/or slab-face sanding processes can get pushed into any and all cracks in a core, including induced fractures, where it resembles a fine white mineralization (Figure D1e.6). Because of the fine grain size of the powder, if the host rock is a carbonate the powder will effervesce more readily than the surrounding rock, and, since it is microscopic and hard to characterize with a hand lens, it can be mistaken for calcareous mineralization. Usually it is concentrated along a crack near the surface of the core.

Figure D1e.1 Efflorescence on a fracture face can resemble mineralization, but if it also occurs on the outer surface of the core, the slab face, or the inside surfaces of the holes made by taking plugs, then the apparent mineralization is not present in the subsurface. This piece of core from a chalk reservoir shows evaporitic rosettes and clusters, possibly gypsum, on a plane that could be mistaken for a mineralized natural fracture, but that is in fact a slab face. Fragments of a vertical, 4-inch diameter core; uphole is unknown.

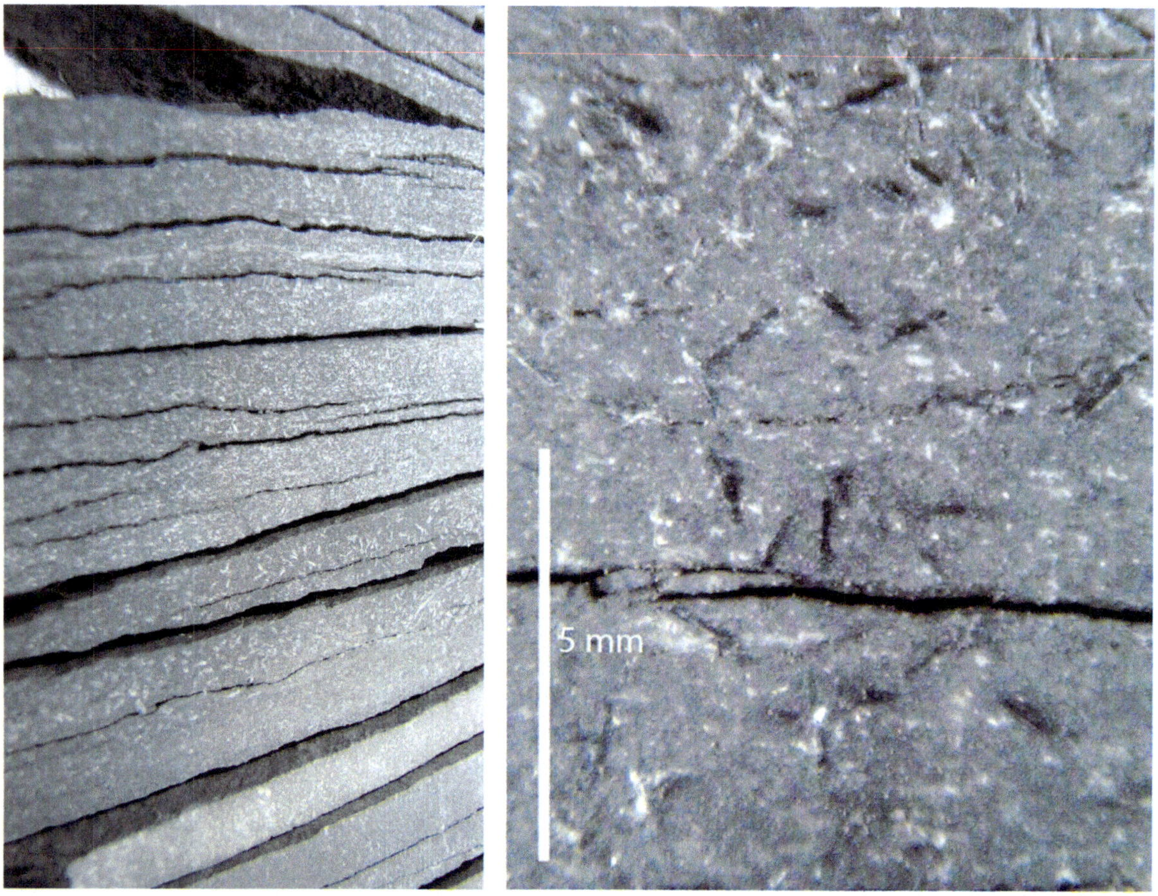

Figure D1e.2 Two views of the small needles of an unidentified evaporite mineral coating the slab face of this core cut from a marine shale. The needles occur on natural fracture faces but also form raised bumps on the slab face and thus must postdate slabbing. Slabs of a vertical, 4-inch diameter core; uphole is towards the top of both photos.

Figure D1e.3 Bivalve shells, particularly but not exclusively in Cretaceous shales, form core-scale partings composed of parallel, prismatic calcite crystals. The shells resemble horizontal "beef"-filled fractures. Criteria for differentiation include bedding plane exposures which may show the growth rings and sometimes remnants of the organic shell linings, and cross-sections which, in the natural fractures, typically have a double layer habit with a medial line. Vertical, 3-inch diameter core; uphole is towards the top of the photo.

Figure D1e.4 Some service companies, or some technicians unknown to the service companies, glue cores together across fractures in order to facilitate handling, particularly slabbing. Although this is usually apparent where the fracture face is exposed, as in these two examples, it is less obvious where the glue has held and the glued fracture plane is exposed only in cross-section. Both photos are of fragments of vertical, 4-inch diameter core, from different marine shale formations.

Figure D1e.5 If a fracture has been cut at an oblique angle by the slab plane, the unsupported wedge of rock on one side of the fracture is subject to cracking and damage during the slabbing and slab-face sanding processes. The shattered, white, cracked rock can resemble mineralization. These three views of a piece of core containing three natural narrow, parallel, high-angle extension fractures show how the middle fracture was badly damaged by the slabbing process, creating a band of what appears to be white mineralization but which consists of cracked and shattered rock. The damage is restricted to the slab face. Slab of a 4-inch diameter core; uphole is towards the top of the upper and middle photos, and away from the viewer in the lower photo.

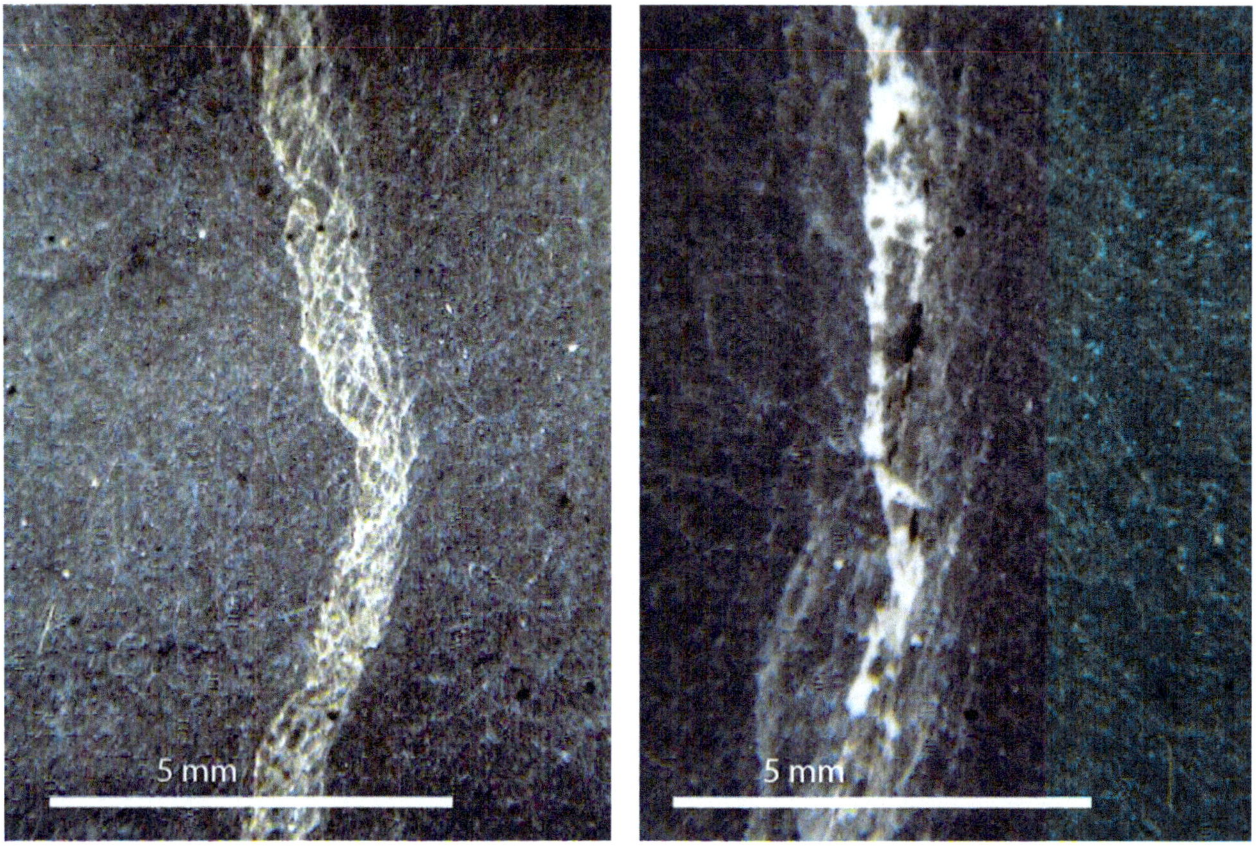

Figure D1e.6 Close-up views of the shattered rock occurring as damage along an induced fracture (left), and of white rock flour "filling" of a damaged, induced fracture (right), on slab surfaces. The vertical fracture planes both strike nearly parallel to the plane of the photos, angling into the photos towards the right. The wedge of unsupported rock between the fracture plane and the slab plane was cracked and damaged during slabbing and slab-face sanding, but remained largely in place. Rock flour created by these processes was trapped in the damage zone. Both the rock flour and the damage zone can resemble mineralization. The faint curlicue lines scratched on the slab surfaces were created by an orbital sanding machine when the slab plane was sanded in order to remove marks left by the slab saw. Vertical core; uphole is towards the top of both photos.

References

Gretener, P.E., 1977, Pore pressure: fundamentals, general ramifications, and implications for structural geology (revised 1979). AAPG Education Course Note Series 4.

Griggs, D., and Handin, J., 1960, Observations on fracture and a hypothesis of earthquakes. Geological Society of America, Memoir, 79, 347–364.

Lorenz, J.C., Warpinski, N.R., Branagan, P.T., and Sattler, A.R., 1989, Fracture characteristics and reservoir behavior in stress-sensitive fracture systems in flat-lying formations. Journal of Petroleum Technology, 41, 614–622.

Lorenz, J.C., Billingsley, R.L., and Evans, L.W., 1998, Permeability reduction by pyrobitumen, mineralization, and stress along large natural fractures in sandstones at 18,300-ft depth: destruction of a reservoir. SPE Reservoir Evaluation and Engineering, 1, 52–56.

Lorenz, J.C., Krystinik, L.F., and Mroz, T.H., 2005, Shear reactivation of fractures in deep Frontier sandstones: evidence from horizontal wells in the Table Rock Field, Wyoming, *in* Bishop, M.G., et al., eds, Gas in Low Permeability Reservoirs of the Rocky Mountain Region: Rocky Mountain Association of Geologists guidebook, pp. 267–288.

Wennberg, O.P., Casini, G., Jonoud, S., and Peacock, D.C.P., 2016, The characteristics of open fractures in carbonate reservoirs and their impact on fluid flow: a discussion. Petroleum Geoscience, 22, 91–104.

Part 2

Induced Fractures

2A

Introduction

Fractures are also created by the coring, core-orientation, and core-handling processes. These non-natural or induced fractures can be common in a core and must be distinguished from natural fractures since they contribute nothing to reservoir permeability and should not be incorporated into a reservoir fracture data base or a fracture-permeability model. Some induced fractures provide stress information and orientation references that can be invaluable in calculating the strikes of natural fractures and other features in the core.

The distinctions between natural and induced fractures are not always obvious. Misleading types of false mineralization can make induced fractures appear to be natural, and induced fracturing can exploit and extend natural fractures, resulting in composite structures with the characteristics of both. In addition, although they are presented here in distinct categories, the range in geometries for induced fracture types overlap; for example, petal fractures that form due to the weight of the drill bit on a formation are commonly extended down the axis of a core to become a single petal-centerline structure when drilling mud exploits the petal fracture cracks to produce a small hydraulic fracture below the bit. Under different stress conditions and forming the other end of the petal-fracture spectrum, petal fractures can form with a saddle geometry that cuts across the axis of a core and overlaps with the characteristics of disc fractures. Similarly, helical torque fractures formed by twisting a core overlap with the jamming fractures created by inadvertently trying to put more rock into a core barrel than it will hold.

The common induced fractures in vertical and horizontal cores that are described here include:

- petal fractures
- saddle fractures
- centerline fractures
- disc fractures
- scribe-knife fractures
- torque fractures
- percussion fractures
- core-jamming fractures
- core-bending fractures.

Of this list, the petal and centerline fractures are perhaps the most easily recognized and the most common. They are also the most useful types of induced fracturing since they almost invariably strike parallel to the in situ maximum horizontal compressive stress. Because of this, they provide a useful and important core orientation reference. Neither of these two induced fracture types is common in cores cut from inclined wellbores since drill-bit stresses are oblique rather than parallel to the in situ formation stresses, but bedding and the wellbore deviation survey often provide enough information to roughly orient inclined cores and any fractures they contain. The other listed induced fracture types do not offer this stress information, but some provide insights on the coring and handling process, and all need to be recognized as induced and therefore not part of the reservoir fracture-permeability system.

Several additional miscellaneous induced fracture types will also be illustrated here, as well as the fractures created by hydraulic stimulation fractures where they have been cored. We also illustrate a unique suite of fractures created by waterflooding.

Important lists of the characteristics that are useful in differentiating natural from induced fractures in core have been published (e.g., Kulander et al., 1990). We offer an abbreviated list below, but it is important to recognize that the characteristics are not universally applicable to each induced or natural fracture type. There are so many caveats and exceptions to most of the criteria that it is usually better for a logger to have a mental image of the characteristics of the various common induced and natural fracture types than to try to use an ambiguous list of conditional characteristics when trying to identify and differentiate fractures.

Atlas of Natural and Induced Fractures in Core, First Edition. John C. Lorenz and Scott P. Cooper.

Nevertheless, ambiguous fractures are common in cores, and lists of recognition characteristics can still be useful and can be a partial substitute for experience. When using a list, a core logger should use multiple, mutually-supporting criteria wherever possible rather than making an interpretation based on single characteristics, and consider the characteristics of the whole fracture set rather than of single fractures wherever possible. This may mean making tentative, preliminary identifications of the first few logged fractures, and supporting or changing that interpretation as more fractures of the set are assessed. As always, one should use all the available core, slabs and butts together, if possible, and the larger pieces of core from rubble zones should be reassembled in order to get the largest possible sample of individual fracture surfaces. Moreover, core segments should be locked together end to end in order to make sure that successive fractures in the core are indeed parallel and part of one set, and to characterize as many of the fractures of that set as possible.

Criteria for Distinguishing Natural from Induced Fractures in Core

Commonly characteristic of induced fractures:

- rough, unmineralized, fresh breaks
- lips at the core edge
- plumes that interact with the core edge and that follow a core axis
- fracture planes that are consistently normal or parallel to the core axis.

Commonly characteristic of natural fractures:

- mineralization
- similar orientations and geometries to mineralized fractures
- no interaction with the core surface
- plumes, steps, or slickenlines that have axes that are unrelated to the core axes
- generally more planar and more systematic than induced fractures.

Reference

Kulander, B.R., Dean, S.L., and Ward, B.J., 1990, Fractured Core Analysis: Interpretation, Logging and Use of Natural and Induced Fractures in Core. AAPG Methods in Exploration Series 8. Tulsa: American Association of Petroleum Geologists.

2B

Petal and Saddle Fractures

Petal fractures may be the most easily recognized type of induced fractures, yet they probably also have the widest range of shapes and sizes (Figure 2B.1). Petals may occur as isolated fractures (Figures 2B.2, 2B.3) or they may be closely spaced and nested (Figures 2B.4–2B.6). They may form on only one side of the core, or they may form as either opposing or staggered fracture pairs on opposite sides of a core (Figure 2B.7).

Petal fracture surfaces may be rough (Figure 2B.8), smooth (see Figure 2B.2), or ornamented by either concave-upward ribs, indicating incremental downhole propagation (Figures 2B.9, 2B.10), or by a plume suggesting more rapid, single-event downhole propagation (Figure 2B.11). Petal fracture planes may extend only a few centimeters into the core or they may extend into centerline fractures and curve to follow the core axis for tens of feet.

Petal fracture dip angles most commonly increase with depth to form a geometry that is concave downwards and towards the nearest core surface (see Figure 2B.11), but some petal fractures are planar or even concave upward (Figure 2B.12). Petal fractures can form with spacings that are so small that they give the core a shattered and hackly appearance, obscuring the petal geometry (Figure 2B.13).

Despite this variability, petal fractures have a suite of common characteristics that can be used for recognition, including a common 180° symmetry across the core axis. Petal fractures in vertical cores typically do not extend from the core surface more than halfway across the core before either terminating blindly or curving downward to follow the core axis. Petal fracture dip angles at a core edge are also always inclined downward (albeit sometimes with a very shallow dip angle) such that the intersection with the core surface forms a rainbow shape when the core is held in its in situ position and the fracture is viewed normal to strike. If a fracture forms a smile in this position, either it is not a petal fracture or the core is being held upside down even if the service company markings indicate otherwise.

Many petal fractures are mere cracks in the rock and hard to see, especially on a rough outer core surface. These cracks are usually more apparent on a slab face provided the slab surface was cut normal or nearly so to the petal fracture strike. Slabbed petal fracture cracks are often stuffed with rock flour generated by the slabbing and sanding processes, and which can resemble mineralization. Other petal fractures completely break the rock apart to reveal the unmineralized induced fracture faces.

Petal fractures form below the bit during coring as the weight of the bit and drill string, of the order of 10 000–15 000 lb, hammers the rock (see Kulander et al., 1990; Lorenz et al., 1990), each stress event producing one fracture or fracture pair. The extension fracture surfaces follow the curved stress trajectory produced by the weight on the bit. Petal fractures may form rapidly, or they may form incrementally as the weight of the drilling mud and pressure pulses from the mud pump crack the rock below the bit. Lorenz et al. (1990) suggested that rotational shear stress between the bit and the rock might alter petal fracture strikes, but this has not been borne out by later observations and the strikes of petal fractures usually reliably follow the trend of the maximum in situ horizontal compressive stress.

It is not clear that all of the features that have been called petal fractures in image logs have the same structural origin as the features found in cores. If they are, they should form concave-upward patterns in the wellbore walls to complement the concave-downward patterns found in cores.

Since petal fractures in a core typically have the same strike within a few degrees, the strike of nearby natural fractures can be estimated based on their intersection angles with petal fractures if the local stress field is known. Even if the stress field is not known, the strikes of natural fractures relative to petal fractures can be used to determine the number of sets of natural fractures in the subsurface as well as their trend relative to the in situ stresses. Fractures striking normal to petal fractures are

Atlas of Natural and Induced Fractures in Core, First Edition. John C. Lorenz and Scott P. Cooper.

also oriented normal to the maximum horizontal compressive stress and may be subject to closure as fluids are withdrawn from fracture apertures during production; fractures oblique to the petals are critically stressed and will be subject to shear.

Petal fractures also occur in horizontal cores where they typically cut abruptly across the core axis to form a concave-uphole saddle shape (Figures 2B.14, 2B.15) since the stress trajectory created by the weight on bit acts at right angles to the overburden stress rather than parallel to it as it does in a vertical hole. These saddle fractures in any given horizontal core have consistent orientations, with the petal lips typically found on the high and low sides of the core. However, the relationship between fracture geometry and the in situ stresses, and the implications of the fracture orientations, have not yet been investigated. Saddle fractures have the same fractographic variations found in petal fractures in vertical cores, with smooth, plumed, or ribbed faces.

Petal and saddle fractures follow the stress trajectories created by the weight of the core bit, which are commonly interrupted by pre-existing natural fracture planes. Petal fractures may cut nearly normal (Figure 2B.16) or nearly parallel (Figure 2B.17) to a natural fracture. They may also follow and exploit a natural fracture plane (Figure 2B.18).

Many cores contain features that have the same general geometry as petal fractures but that are not, in fact, petal fractures. Misleading and false petal-type structures include bedding planes that lip out of a horizontal core, bending tangs described below, and handling-induced breaks that join a natural fracture to the core surface (Figure 2B.19).

Figure 2B.1 Petal fractures in core typically form concave-downward rainbow arcs that cut only a short distance into the core before terminating (left). They may also steepen downward and extend a short distance along the core axis (right). Vertical, 4-inch diameter cores from a marine limestone (left) and a marine shale (right); uphole is towards the top of both photos.

Figure 2B.2 Petal fractures may occur as widely spaced fractures in cracked but otherwise intact core. The core on the left, from an eolian sandstone, was slabbed exactly in half rather than the more common 1/3–2/3 configuration, and the abrupt shift in bedding dip suggests an inconsistency as to which core half was put into the box; petal fracture #1 has the same strike as petal #2, but reassembly of both core halves shows that they have opposing dip azimuths. The smoothly curving fractures transition to more irregular fractures where the petal fractures were extended during core processing and handling. The core on the right, cut from a chalk, has parted along the petal fracture, exposing a smooth, unornamented, isolated fracture face with the downward-steepening dip angle that is a defining although not universal characteristic of petal fractures. Vertical cores, 4-inch diameter (left) and 2-inch diameter (right); uphole is towards the top of both photos.

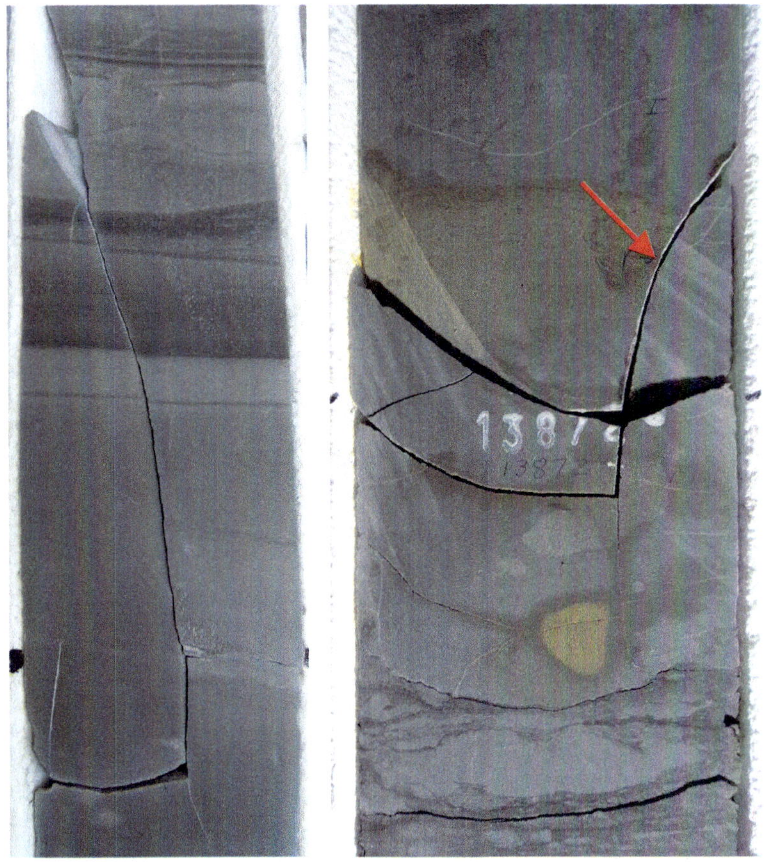

Figure 2B.3 A well-defined, downward-steepening petal fracture (right, arrow) in core from a marine mudstone, and a less regular petal from a marine siltstone (left). Petal fractures commonly terminate downward blindly within a homogeneous lithology, the termination point providing the initiation for later, more irregular core breakage during handling and processing as shown by the cross-core breaks in these two examples.

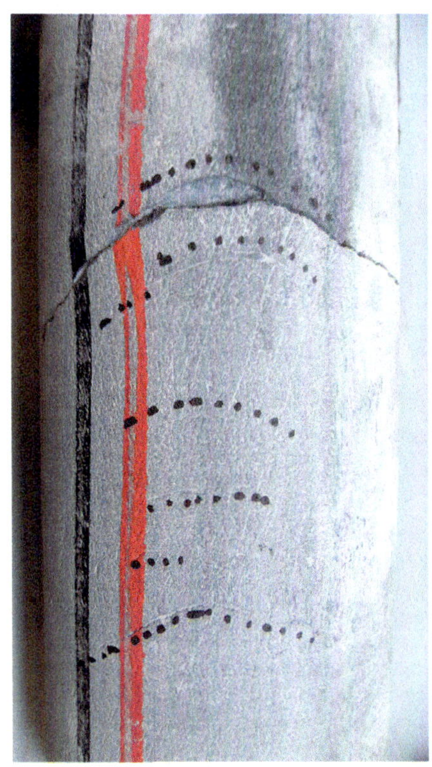

Figure 2B.4 Petal fractures may form subtle, easily overlooked cracks in the rock, recognizable only after washing the drilling mud completely off the core surface and close examination. At least seven parallel petal fracture cracks are outlined by dotted markings on the core. The core has parted along a break formed by the uppermost fracture, which extended across the core as a handling fracture. Vertical, 4-inch diameter core; the red-black line pair indicates that uphole is towards the top of the photo.

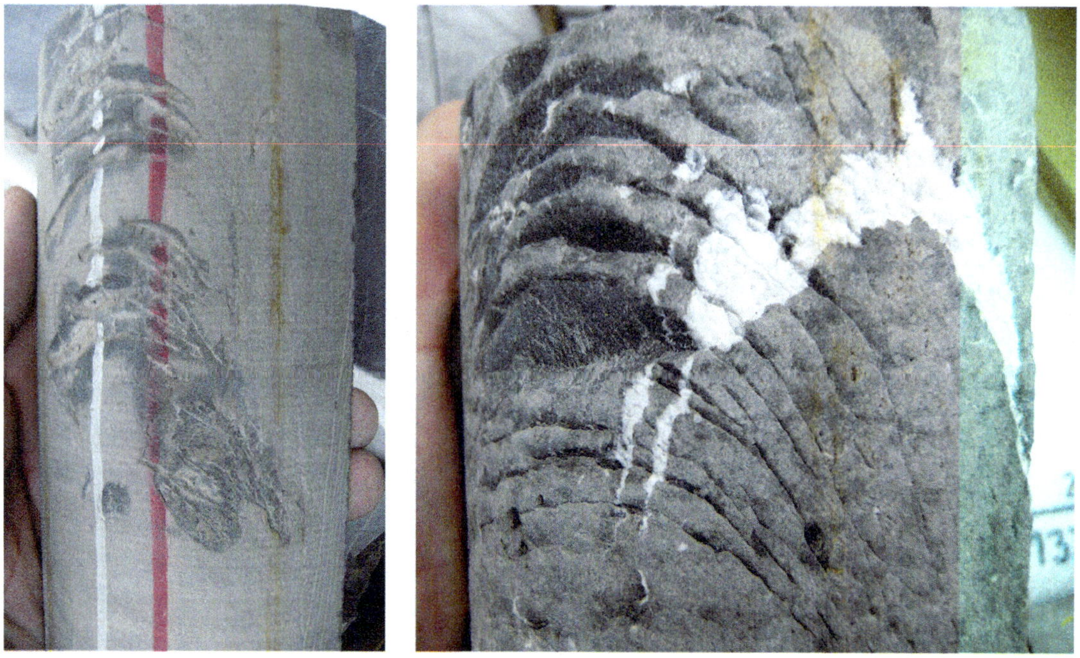

Figure 2B.5 Some petals form very closely spaced fracture systems, shattering the rock but leaving it intact since the petals crack but do not always separate the rock and since they do not cut completely across the core. Slabbing through these systems, depending on the angle of the slab face to the petals, can turn the core to rubble, especially if the opposite side of the core is also shattered by nested petals. These cores are from different marine limestones; the red and white lines drawn on the core on the left are different from the more standard red and black lines but were also used to indicate the up direction on the core, with "red on the right looking uphole." In the right photograph, note the irregular, calcite-mineralized natural fracture that strikes at 90° to the nested petal fractures. This natural fracture must therefore strike 90° to the present-day maximum horizontal compressive stress. Vertical, 4-inch diameter cores; uphole is towards the top of both photos.

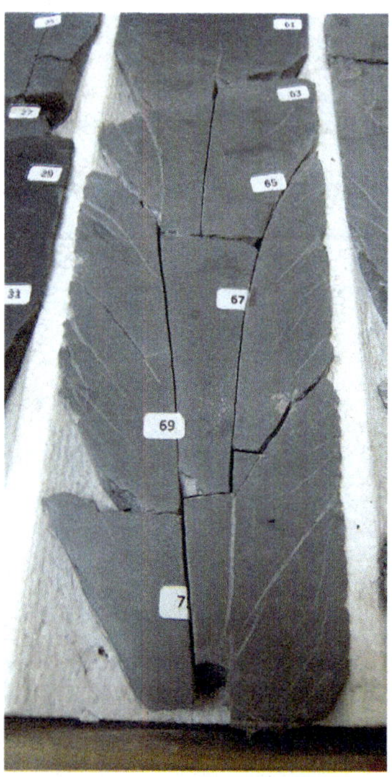

Figure 2B.6 Some nested petal fracture systems are well developed to the point where it is surprising that the core retained any integrity during slabbing. A few of these petal fractures, formed in a marine limestone, extend downward along the core axis, others extend into the core only a short distance before terminating. Although centerline fractures commonly originate from a petal fracture, once the centerline has formed any secondary petals typically approach but do not intersect it; most apparent intersections are the result of core breaks created during processing. Slab from a vertical, 4-inch diameter core; uphole is towards the top of the photo.

Figure 2B.7 Many petal fractures form as fracture pairs with parallel strikes and opposing dip azimuths on opposite sides of the core. The subtle petal fractures in this core, cut from a non-marine limestone-mudstone sequence, are outlined by dotted lines drawn on the slab surface to highlight them for the photo. (Indelible lines drawn on a slab face should be drawn only on the butt piece of the core.) The white material in these petal fractures is rock flour created during slabbing and polishing and is not mineralization. These fractures would have been missed completely if the slab plane had been cut parallel to the petal fracture strikes rather than at 90° to them. Slab from a vertical, 4-inch diameter core; uphole is towards the top of the photo.

Figure 2B.8 Petal fractures surfaces can be quite rough in heterogeneous strata if the stress differential created by the weight on bit is low. This rough petal fracture in a limestone-shale sequence was extended downward as a centerline fracture and is irregular yet can still be recognized by the characteristic increase in dip angle with depth. Vertical, 4-inch diameter core; uphole is towards the top of the photo. The service company used a red-green rather than the more common red-black line pair to indicate uphole/downhole orientation.

Figure 2B.9 Petal fractures such as this one may be ornamented with concave-upward fracture-propagation arrest lines or ribs, suggesting incremental downward growth of the fracture. The closely spaced arrest lines suggest a low core-bit rate of penetration while coring this calcareous siltstone. Vertical, 4-inch diameter core; uphole is towards the top of the photo.

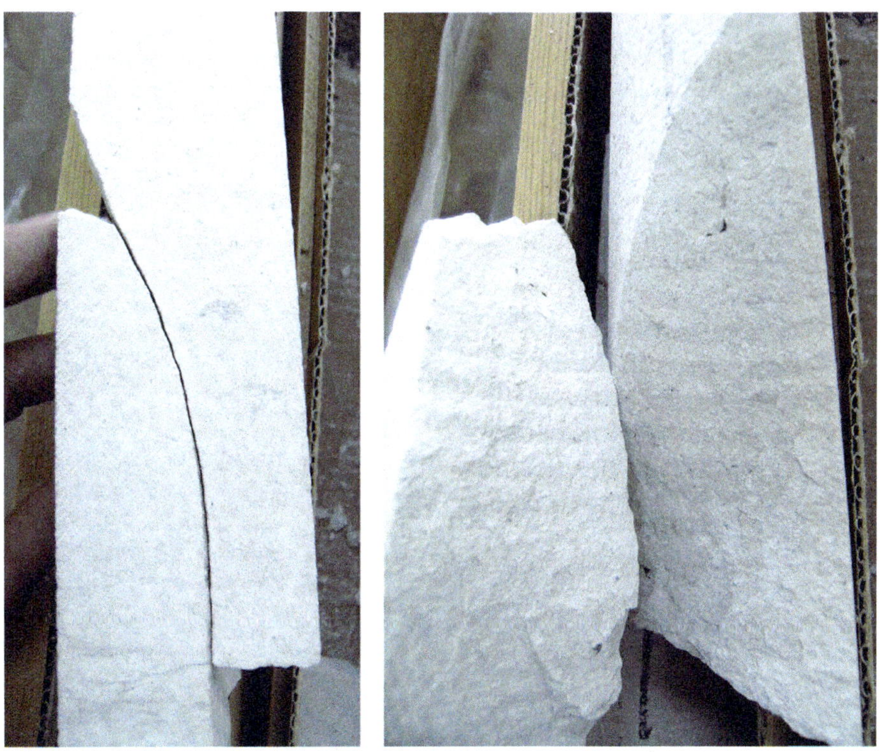

Figure 2B.10 Two views of a petal fracture in a chalk, edge-on showing the downward increase in dip angle (left) and face on, showing broad, concave-upward ribs. The widely spaced ribs suggest that each increment of fracture propagation was of the order of a centimeter and that the rate of penetration while coring this chalk was relatively high. Quarter-section of vertical, 4-inch diameter core; uphole is towards the top of the photo.

Figure 2B.11 Two views of a steeply dipping petal fracture in a limestone, edge-on (left, showing the increasing dip angle with depth), and face-on (right). The well-developed plume structure with an axis parallel to the core axis indicates downward fracture propagation. The plume and the absence of arrest lines suggest that the fracture formed rapidly during a single fracturing event. The subtle fractography of the fracture face has been accentuated by holding it at a low angle to sunlight. Vertical, 4-inch diameter core; uphole is towards the top of the photo.

Figure 2B.12 Petal fractures do not always form steepening-downward structures. If the weight on bit is relatively low and the horizontal in situ compressive stresses are high, petals may have shallow dip angles and may even cut across the core. The left photo shows paired petal fractures (arrows) on opposite sides of a well-cemented sandstone core cut from a sharply folded anticline. The right photo shows core from the same well, with two low-angle, parallel petals on one side of the core, 4 feet deeper. Locking the core together showed that all four petals have the same strike. Vertical, 4-inch diameter core; uphole is towards the top of both photos.

Figure 2B.13 The petal fractures in this small-diameter core cut from a dense, fine-grained limestone are very closely spaced and have shallow dip angles. The left photo shows the subtle cracks in the rock forming rainbows and half-rainbows on one side of the core (the same pattern is repeated on the opposite side of the core but is not present at the 90° left or 90° right positions on the core surface). The right photo, of the bottom of the same core piece looking upward, shows the parallel-striking (black lines), shallow-dipping fracture faces. Vertical, 2.5-inch diameter core; uphole is towards the top of the upper photo, and away from the viewer in the lower photo.

Figure 2B.14 Joined, low-amplitude petal fractures at the high and low sides of this horizontal core form a saddle with a smooth surface. The two scribe-line grooves engraved on the core indicate that the core is oriented (horizontal cores are oriented relative to up rather than to north). Horizontal, $2^{9/16}$-inch diameter core; stratigraphic up is towards the top of the photo, uphole (toward the heel of the well) is to the right.

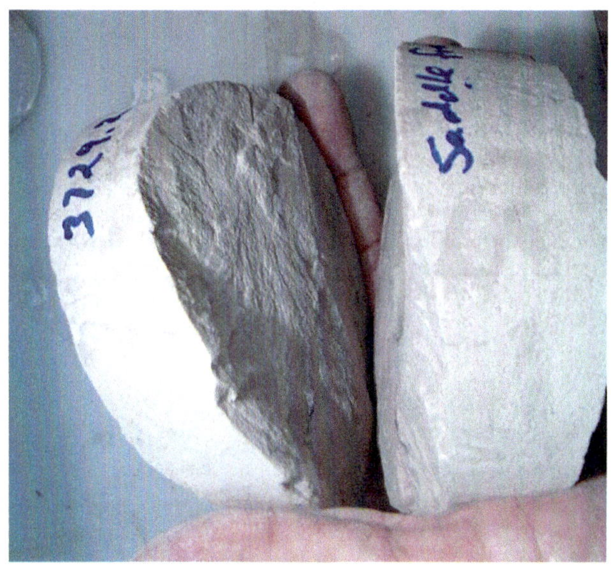

Figure 2B.15 A shallow saddle fracture in horizontal core from a dense chalk, marked by plume structure that originates at the tip of one low-amplitude petal. Horizontal, 4-inch diameter core; stratigraphic up is towards the top of the photo, uphole is to the left.

Figure 2B.16 Petal fractures ("P", adjacent to the dashed black lines drawn on the core) that strike at high angles to a natural fracture ("N") are commonly discontinuous across the mechanical discontinuity created by the natural fracture (upper petal fracture), or they may terminate against the natural fracture (lower petal fracture). The core surface is grooved by scribe knives and marred by wider scars caused by a dragging core catcher. Vertical, 4-inch diameter core; uphole is towards the top of the photo.

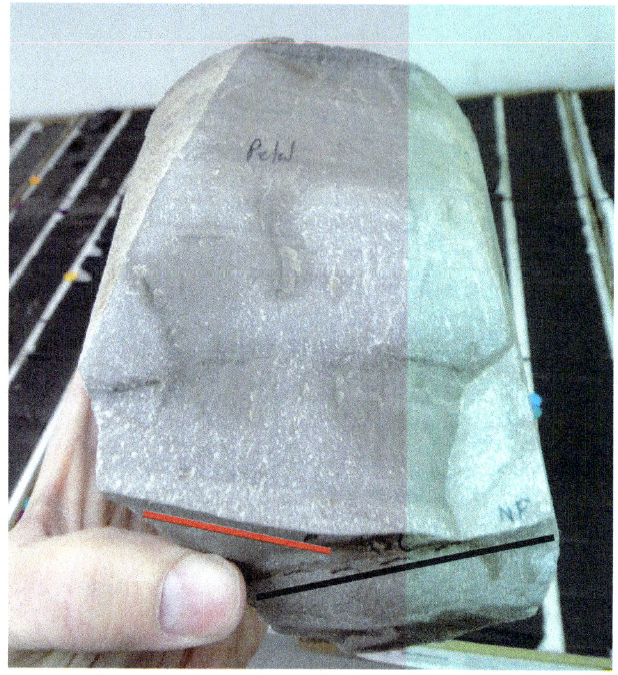

Figure 2B.17 A petal fracture, parallel to the red line, formed in core cut from a marine limestone strikes at a 35° angle to a calcite-mineralized natural fracture ("NF") and terminates against the discontinuity offered by the natural fracture. The surface of the petal fracture has a patchy, white cast similar to the calcite that coats the natural fracture face, but the surface of the petal fracture consists of damaged rock that merely resembles mineralization, a common occurrence in fine-grained, dense limestones. Vertical, 4-inch diameter core; uphole is towards the top of the photo.

Figure 2B.18 A composite natural and induced fracture system in core cut from a non-marine shale-limestone sequence. Surfaces B and D are ornamented with plumes that propagated horizontally across the fracture face, indicating that it is a natural fracture, supported by the light coating of calcite mineralization on the fracture face at C where it intersects a limestone bed. In contrast, surface A is an unmineralized, steepening-downward petal fracture that strikes parallel to the natural fracture but that doesn't quite join it, as indicated by the step of broken rock at B. Vertical, 4-inch diameter core; uphole is towards the top of all photos.

Figure 2B.19 Petal fractures can share a general morphology with other structures which must not be mistakenly identified as petal fractures. The two photos shown here show two views of a feature that might be logged as a petal fracture except that (1) according to the red-white line pair it would be an upside-down petal, and (2) when viewed with oblique lighting (right), the more planar section of the structure displays the horizontally propagating plume structure and unmineralized but aged surface of a natural fracture. Moreover, the natural fracture surface contrasts sharply with the induced, curved lip that joins the edge of the natural fracture (white arrow) at the bottom of the structure to the core surface. The core was slabbed normal to the natural fracture. Vertical, 4-inch diameter core; uphole is towards the top of the photo.

References

Kulander, B.R., Dean, S.L., and Ward, B.J., 1990, Fractured Core Analysis: Interpretation, Logging and Use of Natural and Induced Fractures in Core. AAPG Methods in Exploration Series 8. Tulsa: American Association of Petroleum Geologists.

Lorenz, J.C., Finley, S.J., and Warpinski, N.R., 1990, Significance of coring-induced fractures in Mesaverde core, northwestern Colorado: American Association of Petroleum Geologists Bulletin, 74, 1017–1029.

2C

Centerline Fractures

Induced centerline fractures typically split a core approximately in half, and can extend anywhere from a few centimeters to many meters along a core axis (Figures 2C.1, 2C.2). Some underpressured formations are particularly susceptible to centerline fracturing with cores containing multiple parallel centerlines (Figures 2C.3, 2C.4) whereas these structures are less well developed or even absent in cores cut from other formations. Many cores cut from resource-play shales are split by a variably continuous centerline fracture for nearly the full length of the core.

The ribbed surfaces of many centerline fractures (Figures 2C.5, 2C.6) suggest an origin as small hydraulic fractures below the core bit, propagated by the weight of the drilling mud and the pressure pulses from the mud pump. In fact, the ribs found on some centerline fractures show a four-fold cyclicity in amplitude and spacing, possibly related to mud pump cycles (Figure 2C.7). Other centerline fracture surfaces have fractographic markings that suggest long intervals formed in single, rapid events then propagating in short incremental intervals (Figure 2C.8), and some centerlines are rough and highly irregular (Figure 2C.9).

Centerlines commonly originate from petal fractures (Figure 2C.10), and in fact, Kulander et al. (1990) treated petals and centerlines as single, combined, petal-centerline structures. However, we have found centerlines that originate blindly in the middle or at the side of a core without associated petals (Figure 2C.11) while isolated petals are common, and the two can form as separate, stand-alone structures. Centerline fractures can also form in an uncored wellbore, and are probably equivalent to the drilling-induced fractures recorded by image logs.

As noted by Kulander et al., the radius of the arrest-line ribs that are common on many centerline surfaces suggests that they should be large enough to extend beyond the core radius and into the walls of a wellbore. However, the edges of centerline fractures are occasionally captured by cores and show that the propagating edges of centerline fractures do not have uniform radii. Rather, the edge of a centerline fracture bends upward abruptly at the edge of the fracture, giving it a much smaller lateral dimension than would be projected by the assumption of a uniform radius (see Figure 2C.7). Although the fractures may extend for many feet along a core and along the equivalent section of a wellbore, most probably do not extend more than a few inches into the adjacent formation.

Centerline fractures commonly cut indiscriminately across minor bedding boundaries, suggesting that propagation is being driven by hydraulic forces that are strong relative to minor lithologic, mechanical property differences. Centerlines commonly terminate downhole blindly in a core for no apparent reason (Figures 2C.12, 2C.13). In many cores, centerline fractures terminate at or just above the depth at which a new petal-centerline fracture begins (Figure 2C.14), suggesting that the new fracture takes over the burden of stress accommodation. In other cores, a medial centerline fracture may cut through a suite of petal fractures that curve from one or both sides of the core to become tangent and parallel to the centerline, generally terminating before quite intersecting it. Other cores can be cut by numerous petal fractures and parallel centerlines (see Figures 2C.3, 2C.4). Some centerlines terminate at faults and other heterogeneities, with twist hackle on the fracture plane recording the altered stress conditions at the mechanical heterogeneity (Figure 2C.15).

Centerlines typically deviate around concretions and other larger heterogeneities in a formation. They may follow natural fractures for short distances or cut across them (Figures 2C.16–2C.18), but centerline fracture propagation appears to be determined more by the stress anisotropy in a formation than by the mechanical heterogeneities provided by small mineralized fractures. Nevertheless, centerlines are less common in cores containing large natural fractures with open apertures which divert the fluids and pressures that drive centerline fracturing.

Oriented cores show that centerline fractures strike parallel to the maximum horizontal in situ compressive stress, as would be expected of hydraulic fractures. As with petal fractures, they provide an important orientation reference for logging the approximate natural

Atlas of Natural and Induced Fractures in Core, First Edition. John C. Lorenz and Scott P. Cooper.

fracture strikes in a core, either true strikes if the trend of the in situ stress is known, or at least the strikes of the fractures relative to each other and to the in situ stress if the stress trend is not known.

A relatively rare type of induced centerline fracture forms as scattered, irregular vertical extension structures roughly parallel to the axis of a vertical core (Figures 2C.19, 2C.20). The core may be incompletely separated along the cracks. If the core is deliberately cracked open along these cracks, however, the highly irregular fracture faces are revealed as freshly broken rock (Figure 2C.21).

Good examples of such fractures were noted in the lower 30 ft of a 121 ft-long core. The core had been cut with a core barrel that was only 120 ft long, a foot shorter than the cored interval. As the excess foot of core was forced into the bottom of the barrel, some of the weight of the drill string was supported by the core which had reached the limit of travel at the top of the core barrel.

The resulting compressive stress parallel to the core axis turned the core into a large-scale laboratory experiment akin to those reported on 1-inch diameter cylindrical rock samples by Griggs and Handin (1960) or Wawersik and Fairhurst (1970), where extension fractures form parallel to the applied load in laterally unconfined samples. The core was not confined laterally except by the fluid pressure of the drill mud, so it expanded laterally under the load applied parallel to the core axis. These fractures may be related to the induced fractures with curved strikes described in Section 2K.

The lowest foot of core was crushed vertically while being twisted, creating a distinctive rubble consisting of helical torque fractures, as described in section 2 F.

Figure 2C.1 Centerline fractures occur in all cored lithologies, including marine shales (left) and limestones (right). They can extend for tens of feet along a core, wandering but not exiting the core, suggesting that they are controlled by the stresses close beneath the core bit. Centerline fractures are commonly ribbed (left) or rough-surfaced (right), and, more rarely, marked by plumes indicating downhole propagation. Vertical, 4-inch core (left) and slabs from a 4-inch core (right); uphole is towards the top of both photos.

Figure 2C.2 This centerline fracture in a core cut from a calcareous marine shale has an undulating but nearly vertical dip, and splits the core into approximately one-quarter/three-quarter segments rather than halves. One hundred and eighty feet of core were recovered from this well, and 90% of the core was split by four parallel, long centerline fractures. Each of the four successive centerline fractures started as an extension of a petal located immediately below the termination of the overlying centerline. Vertical, 4-inch diameter core; uphole is away from the viewer.

Figure 2C.3 Three, parallel centerline fractures (arrows) formed in core cut from a marine siltstone, cutting the core into board-shaped slabs. Sections with this multiple centerline fracture geometry extend for up to 10 feet along the core. The shiny, metallic flakes are remnants of the inner core barrel after it was sawn lengthwise in order to extricate the core from the barrel. Vertical, 4-inch diameter core; uphole is towards the top of the photo.

Figure 2C.4 Eight parallel centerline fractures (numbered) are present in this section of core cut from a coarse-grained marine limestone. Butt from a vertical, 4-inch diameter core; uphole is towards the top of the photo.

Figure 2C.5 The curvature on some centerline ribs has a small radius, suggesting a limited lateral extent. This small-radius fracture in core cut from a marine shale probably did not extend to the wall of the wellbore. The core captured the right edge of the fracture (red bracket). Fracture propagation jumped from incremental, marked by ribs in the top half of the core, to rapid, marked by a plume in a more brittle bed near the bottom of the photo, then returned to incremental, ribbed propagation where the lithology returned to a clayey shale at the bottom of the photo. Vertical, 4-inch diameter core; uphole is towards the top of the photo.

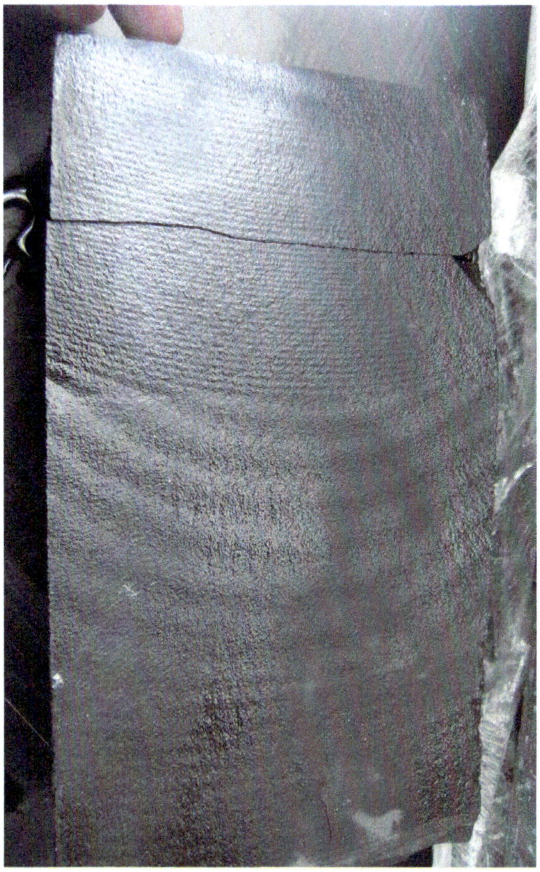

Figure 2C.6 Variable rib spacing and radius of curvature on a centerline fracture face record three rates of fracture propagation through three different lithologies in this marine calcareous shale core. Bedding planes visible on the outer core surface correlate to the changes in rib spacings, but the subtle differences in clay and carbonate content responsible for the variations could not be quantified visually. Section of a vertical, 4-inch diameter core; uphole is towards the top of the photo.

Figure 2C.7 This example, cut from a muddy marine shale, definitively captured the lateral edge of a centerline fracture as shown by the abrupt upturn of the ribs and the unbroken rock to the left of the upturn. The very narrow nature of the centerline crack is also indicated by the spur of rock that extends unbroken across the fracture plane near the base of the photo. The ribs have a cyclic pattern, with a series of several small ribs, commonly three, separated by a large rib and a flat zone, possibly related to a four-fold cyclicity in strokes of the mud pumps on the rig floor. Butt of a vertical, 4-inch diameter core, slabbed face on the right edge; uphole is towards the top of both photos.

Figure 2C.8 Centerline fracture surfaces marked by plume structure (left) suggest rapid propagation of the fracture over this interval of the core. This section of the centerline originated from the curved fracture arrest front shown at the top of the core. Centerline fractures can also be rough and unornamented, or they may be marked with upward-concave ribs/arrest lines (right), suggesting incremental propagation, possibly related to pulses of the mud pump which can effectively add a pound to the weight of the drilling mud density at the bit face. Note that the right edge of the core was chipped by rotation of the core bit around the fractured core as it was pushed into the core barrel; this indicates the fracture propagated in front of the drill bit. The right edge is chipped by spalling of the core edge as the clockwise-rotating bit pulled on it in tension, whereas the left edge, where the rotating bit acted on the rock in compression, is not chipped (see Kulander et al., 1990). Sections of vertical, 4-inch cores; uphole is towards the top of both photos.

Figure 2C.9 Some centerline fractures, especially in muddy shales, can be quite irregular. This centerline fracture, in a core cut from a muddy marine shale, wanders along the middle of the core and is not deflected by the minor lithology changes. The zig-zag, step-wise propagation suggests that the formation overburden stress was not quite aligned with the bit-weight stress. Butts of a vertical, 4-inch diameter core; uphole is towards the top of the photo.

Figure 2C.10 Centerline fractures (red arrows) extending downhole from curved petal fractures, and smaller petal fractures (white arrows) that were not extended into centerlines, in core cut from a siliceous marine shale. Vertical, 4-inch diameter core; uphole is obliquely out of the photograph toward the viewer's right.

Figure 2C.11 Some centerline fractures originate not from a petal fracture but from the side in the center of a core. These two examples, in cores cut from two different calcareous marine shale formations, show broad rib markings recording propagation from the left side of the core, across the core, and then extending a short distance downward along the core axis as recorded by a pattern of finer ribs with twist hackle. Both fractures extend downhole for several feet below the intervals shown in the photos. Vertical, 4-inch diameter cores; uphole is towards the top of both photos.

Figure 2C.12 Centerline fractures may terminate downhole for no apparent reason within a homogeneous lithology. If the core has been handled roughly, the rock may break at the fracture termination, turning abruptly towards the core surface as in this marine shale. Butts of vertical, 4-inch diameter core; uphole is towards the top of the photo.

Figure 2C.13 The ribs on this centerline fracture face show that it propagated a short way into a fine-grained limestone and terminated blindly in the homogeneous rock at a curved arrest line. The semi-planar gray surface in the rock below the ribs is a fresh, induced extension of the centerline fracture created during core handling.

Figure 2C.14 This irregular centerline fracture in a core cut from a calcareous shaley marine limestone terminates blindly in a homogeneous lithology at the same depth that another irregular petal fracture cuts in from the right side of the core. This is a common arrangement, suggesting that both fractures were not required to accommodate the strain in the system. Vertical, 4-inch diameter core; uphole is towards the top of the core.

Figure 2C.15 This centerline fracture in core cut from a siliceous marine shale terminated at the bottom of the photo ("End") at a major change in lithology after breaking into twist hackles, each with its own set of rib markings. Section of a vertical, 4-inch diameter core; uphole is towards the top of the photo.

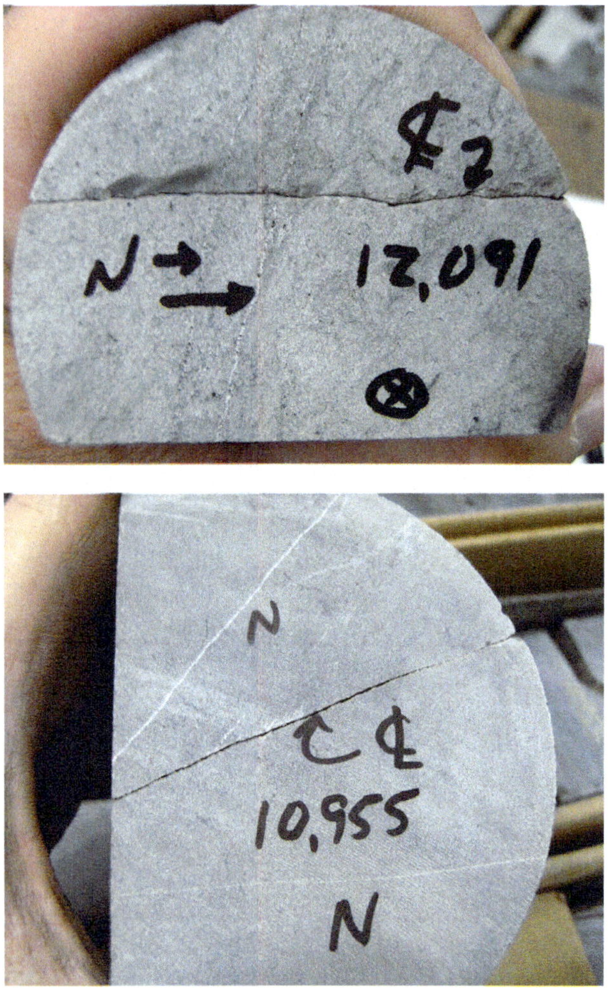

Figure 2C.16 Centerline fractures ("CL") striking oblique to smaller, mineralized natural fractures ("N") may or may not be deflected by or interact with the natural fractures. In the upper example, from a non-marine sandstone, the centerline cuts across two calcite-mineralized fractures without so much as a hiccup. In the lower example, from a muddy non-marine siltstone, the centerline strikes at a low, oblique angle to one of the natural fractures in the core, but did not exploit what was presumably a plane of weakness in the rock. Butts from vertical, 4-inch diameter cores; uphole is away from the viewer in both photos.

Figure 2C.17 A centerline fracture (the surface parallel to the red lines), in core cut from a marine limestone, is offset at a narrow, calcite-mineralized natural fracture (along the white dotted line). The centerline fracture propagated downhole, away from the viewer, and followed the oblique natural fracture for a distance of only a few millimeters, suggesting that the natural fracture is a weakness plane in the rock but that stress was the main parameter controlling fracture orientation, not the natural fracture fabric. Section of a vertical, 4-inch diameter core; the circled dot indicates that uphole is towards the viewer.

Figure 2C.18 Even nearly parallel, poorly mineralized natural fractures may not control the propagation of centerline fractures for any great distance. This core, cut from a calcareous marine shale, shows a ribbed centerline fracture in the top half of the photo that propagated into and exploited an unmineralized, smooth-faced natural fracture in the central part of the photo. The centerline fracture quickly broke out of the natural fracture even though there is only a 5° difference between the two fracture strikes, and continued to propagate down the axis of the core. The bracketed zone indicates an area of overlap with minimal interaction between the centerline and natural fracture. Vertical, 4-inch diameter core; uphole is towards the top of the photo.

Figure 2C.19 A system of short, scattered, irregular, high-angle (vertical) induced extension fractures. What appears to be mineralization is rock flour caught in the cracks. Slabs of 4-inch diameter vertical core; uphole is towards the top of the photo.

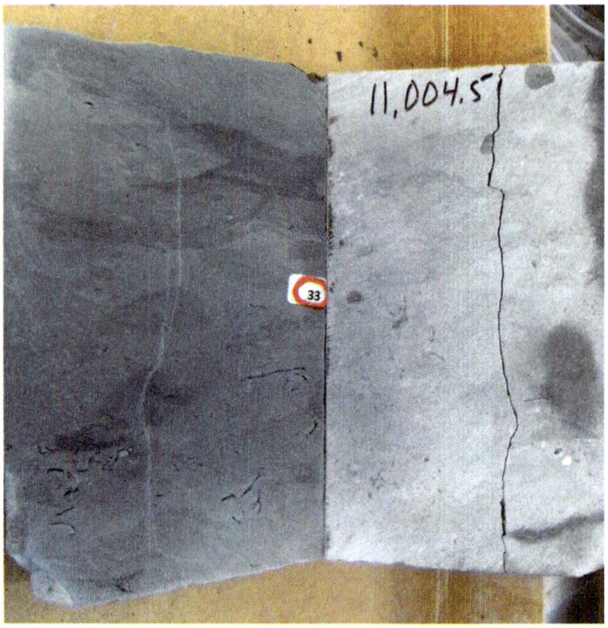

Figure 2C.20 Slab (left) and butt (right) of a 4-inch diameter vertical core, showing an irregular, high-angle, induced extension fracture. The fracture was just a crack in intact core until the butt was intentionally opened along the crack for examination (see the following photo). Uphole is towards the top of the photo.

Figure 2C.21 The highly irregular and fresh surface of the induced extension crack shown in the previous photo. Butt from a 4-inch diameter vertical core; uphole is towards the top of the photo.

References

Griggs, D., and J. Handin, 1960, Observations on fracture and a hypothesis of earthquakes. Geological Society of America, Memoir, 79, 347–364.

Kulander, B.R., Dean, S.L., and Ward, B.J., 1990, Fractured Core Analysis: Interpretation, Logging, and Use of Natural and Induced Fractures in Core. American Association of Petroleum Geologists, Methods in Exploration Series, No. 8. Tulsa: American Association of Petroleum Geologists.

Wawersik, W.R., and C. Fairhurst, 1970, A study of brittle rock fracture in laboratory compression experiments. International Journal of Rock Mechanics and Mining Science, 7, 561–575.

2D

Disc Fractures

Induced disc fractures have long been recognized in the mining industry, being well developed in the short, small-diameter cores cut from the walls of hard-rock mines, and the mining literature contains numerous discussions of their possible origin (e.g., Obert and Stephenson, 1965). Disc fractures are also common in many of the cores cut from oil and gas wells, typically occurring at regular, often short intervals along a core, breaking it into regular slices ranging in thickness from centimeters to tens of centimeters (Figures 2D.1, 2D.2).

Several mechanisms may form repetitive core-normal fractures, and it is not always apparent which mechanism or set of mechanisms applies to a given suite of disc fractures. Although there are discrete disc fracture categories, there appears to be a spectrum of induced fractures, and presumably mechanisms, that grades from disc fractures to saddle fractures. The spectrum may also overlap with certain types of torque fracturing. Some of the mechanisms that have been suggested to explain repetitive fractures oriented normal to the long axis of a core include:

- concentrated stresses at the base of a wellbore during coring due to the weight on the core bit and/or the weight of the drilling fluids (e.g., Li and Schmitt, 1998)
- vertical relaxation and extension of a core after being released from the weight of the overburden, enhanced in shaley rock by the weakness planes along bedding
- slaking (the disintegration of rock during cyclic wetting and drying) due to dehydration of a core, particularly in muddy shales (Figure 2D.3)
- flexure of the core barrel, leading to regular core-normal breaks along the core in the barrel
- gas expansion during rapid core retrieval from the subsurface, leading to parting along bedding planes as gas pressure in the core is released
- shear, as suggested by repetitive core-normal fractures with asymmetrically stepped and lineated surfaces (Obert and Stephenson, 1965) (Figures 2D.4, 2D.5)
- rotational shear at regular intervals caused by spinning one core section against another, producing spinoffs (described in Part 3, Artifacts).

Characteristics such as core-confined plumes (Figures 2D.6–2D.12) suggest that the most common type of disc fracture is formed in extension after the core is cut. Similarly, lips or small cusps at the intersection between disc fractures and outer core surfaces (Figures 2D.7, 2D.8, 2D.10, 2D.12, 2D.13), and plume patterns that curve to become normal to the outer core surface (see Figures 2D.9 and 2D.12) confirm that a free core surface existed at the time of discing. Interactions with natural, induced centerline and induced petal fractures also indicate that this type of extensional disc fracturing postdates these other structures. Plume axes that follow the in situ formation stresses (see Figure 2D.8) and rims of drilling mud that must have invaded the edges of disc fractures while the mud was still wet (Figures 2D.7, 2D.14) indicate that some disc fractures formed at an ill-defined time "soon after" the core was cut.

Some disc fractures, commonly those found in cores cut from poorly bedded silty mudstones, occur as compound structures consisting of an older core-normal plane and a younger, superimposed cone-shaped ring structure (Figures 2D.15, 2D.16). The cones can be either concave-upward or concave-downward, and if viewed only on the two-dimensional slab plane they appear as wedges of rock. Other disc fractures consist of core-normal cracks that don't fully separate the rock across the fracture plane (Figures 2D.17–2D.19).

The term "disc," if defined by geometry alone, covers core-normal fractures that break a core into short cylinders, which happens in both vertical and horizontal cores. Vertical expansion and related cracking also occur in deviated and horizontal cores, forming similar horizontal extension fractures with comparable plumes and edge lips. However, the fractures are not disc-shaped since the core parts parallel to rather than across the core axis. Mechanically if not geometrically, the definition of disc fractures can be expanded to include non-disc-shaped fractures in cores cut from deviated and horizontal wells (Figures 2D.20, 2D.21). In fact, some horizontal cores contain both vertical/core-normal, induced disc fractures created by bit stresses and horizontal extension

Atlas of Natural and Induced Fractures in Core, First Edition. John C. Lorenz and Scott P. Cooper.

disc fractures created by release of vertical stress on the rock. As with other induced fracture types, disc fractures may originate from or exploit the weakness planes provided by natural fractures (Figure 2D.22).

Induced disc fractures do not provide information regarding reservoir permeability, and they must be recognized when logging a core so that they are not included in the conceptual or quantitative models of fracture-controlled fluid flow in a reservoir. Many horizontal disc fractures in vertical cores have been erroneously interpreted to be natural fractures.

Figure 2D.1 Disc fractures that cut a core into "poker chips" or "hockey pucks" (terminology depending on both spacing and the labeler's hobbies) are particularly common in cores cut from bedded shales. The bed-parallel disc fractures in this core, cut from a siliceous marine shale, have average spacings of a few centimeters. These are induced fractures, but that determination cannot be made from this photo; the differentiation between natural and induced fractures with disc geometries should be based on characterizations that include fractography, indications of relative ages, and structural context. Vertical, 4-inch diameter core; uphole is to the right.

Figure 2D.2 Most disc fracturing postdates the formation of centerline fractures, as shown here by the termination of some of the numerous discs against the centerline fracture in this core cut from a marine limestone. The rectangular voids in the core are the result of the different out-of-place positions of disc segments along the core. The core has just been cut from, and is still cradled by, half of the aluminum inner core barrel. Vertical, 4-inch diameter core; uphole is away from the viewer.

Figure 2D.3 Many muddy shale cores become badly broken up by well-developed disc fracturing oriented parallel, subparallel, and oblique to bedding after being cut from the formation, as are these cores cut from a marine interbedded shale-limestone formation (left) and a calcareous shale (right). Disc fractures can continue to form for days, weeks, even months after the core is cut, frustrating the core logger who is trying to obtain fracture counts, and suggesting that dehydration and core relaxation after release from the vertical overburden stress play a role in some types of disc fracturing. Vertical, 4-inch diameter core; uphole is towards the top of the photo.

Figure 2D.4 Asymmetrically stepped surfaces and lineations on some core-normal disc fractures, most frequently found in cores cut from well-cemented sandstones and dense, fine-grained carbonates, support suggestions that some disc-type fractures can form in shear (e.g., Obert and Stephenson, 1965), although the mechanism is unclear and the distinction between them and natural, bed-parallel shear fractures is not always obvious. Lineations on these surfaces trend parallel, and the steps trend normal, to the strikes of associated petal fractures, suggesting that steps and lineations are controlled by the in situ maximum horizontal compressive stress and, where present, can be used as consistent core-orientation references. The occurrence of the fracture shown here as one of a series of closely spaced, parallel horizontal planes, and the fact that the planes occur within a brittle, homogeneous lithology rather than along mechanical weakness planes such as clay partings, suggest that this is an induced fracture and not a natural, bed-parallel, flexural-slip shear plane. Half-slab of a vertical, 2.5-inch diameter core; uphole is away from the viewer.

Figure 2D.5 The linear patterns with scattered asymmetric steps on this core-normal fracture in a dense limestone resemble the more definitive stepped surfaces on stepped disc fractures found in quartzites, as shown in the previous photo. Butts from a vertical, 4-inch diameter core; uphole is towards the top of the upper photo, and away from the viewer in the lower photo. Red arrows point to a correlation point for the two photos.

Figure 2D.6 Disc fracture faces in fine-grained lithologies, particularly shales, are commonly marked by fine plume structures that originate from some inhomogeneity in the rock such as a fossil or concretion. In muddy shales such as this core, the plume can be accentuated, once, by lightly rubbing a finger over the fracture surface. The radius of curvature on such plumes is commensurate with the core diameters, in contrast to the broader radiations formed on plumes from larger natural fractures. Vertical, 4-inch core; uphole is towards the viewer.

Figure 2D.7 Plume structures can also be highlighted by oblique lighting. Kulander et al. (1990) noted that the propagation axes of plume structures on discs, such as the axis highlighted by the dotted line marked on this core cut from a calcareous marine shale, commonly record the orientation of the maximum horizontal compressive in situ stress. Plume axes should be more definitive in stiffer rock and/or where the stress differential is higher; lower differentials or even isotropic stresses, and less biaxial/more radial propagation, would be expected in more ductile, clay-rich shales where the two horizontal stresses are nearly equal. Note the lip on the disc fracture at the outer core surface, indicating interaction between the disc plane and the free surface, and that discing formed after the core was cut. Note also the rim of mud invasion next to the core surface, indicating that the mud coating on the core was still wet and mobile at the time of discing. Vertical, 4-inch diameter core; uphole is towards the viewer.

Figure 2D.8 Most plumes on discs are limited in extent, as is this one that formed in core cut from a calcareous marine shale. The transition of the plume to rougher parts of the fracture face also points to discing after the core had been cut. The same pattern also indicates that discing occurred after the centerline fracture (the planar surface at the top of the photo next to the circled X) had formed. The maximum horizontal compressive stress orientation in the formation, indicated both by the axis of the plume (approximately following the dotted line marked on the core) and by the strike of the centerline fracture, is nearly parallel, as would be expected from the Kulander et al. (1990) contention that disc plume axes follow the maximum in situ horizontal compressive stress. Section of a vertical, 4-inch diameter core; uphole is away from the viewer.

Figure 2D.9 Plume structures that decorate disc fractures commonly radiate out across a fracture so that the lineations of the plume intersect the core surface at nearly right angles, indicating that the core surface was present at the time of fracturing. This plume, originating at a pyrite-filled, ptygmatically folded fracture in a marine shale, propagated across the middle of the core along a linear trend reflecting the in situ stress, but then radiated out to intersect the free core surface at right angles. The small lip at the core surface also indicates the presence of a free surface at the time of fracturing, and the invasion rim of drilling mud indicates the fracture formed early, while the mud was still wet and mobile. Vertical, 4-inch diameter core; uphole is away from the viewer.

Figure 2D.10 Some plumes record fracture propagation in what would appear to have been a nearly isotropic horizontal stress condition. The plume on this disc fracture, in a core cut from a siliceous shale, originated at a fossil fish scale and propagated in nearly radial fashion to the core edge, where it formed a lip at the free surface of the core. The irregular shading in the photo is due to the oblique lighting necessary to highlight the fracture fractography. Vertical, 4-inch diameter core; uphole is towards the viewer.

Figure 2D.11 Plumes show that some disc fractures, such as the discs in the muddy marine shale shown here, are formed from the coalescence of multiple smaller fracture planes. The plumes have been highlighted by lightly rubbing the fracture face with a finger; excessive rubbing destroys the subtle fractography. Butt of a vertical, 4-inch diameter core; uphole is towards the viewer.

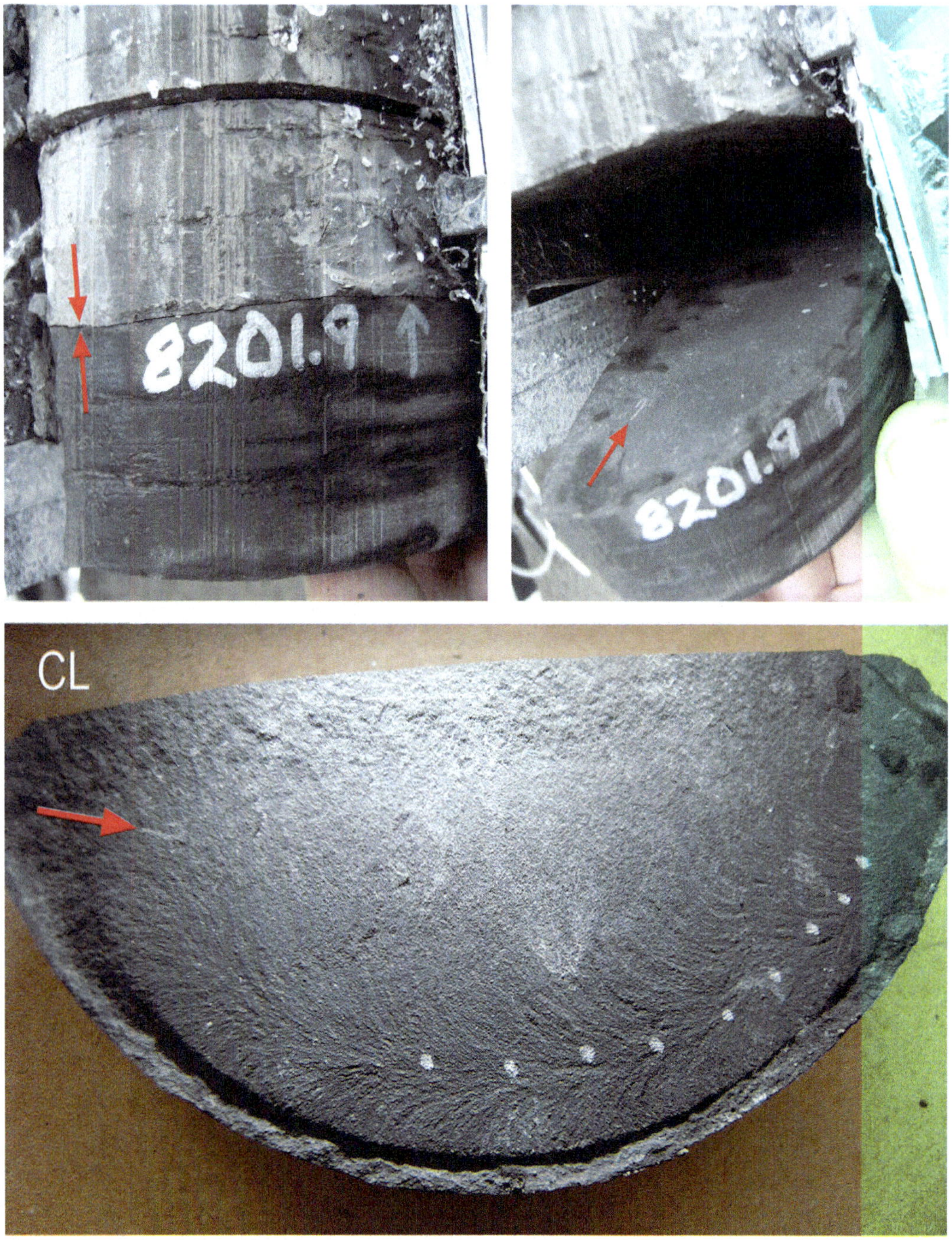

Figure 2D.12 Three views of a subtle disc fracture in a marine limestone-shale sequence (between the arrows, upper left photo) that displays a core-edge lip, a curved plume axis that follows the outer core surface, and a misty zone adjacent to the centerline fracture ("CL"), indicting that discing after the core was cut and after the centerline fracture had formed. For correlation, the red arrows in upper right and lower photographs point to the same hook-shaped scratch on the fracture surface. Butts of vertical, 4-inch diameter core; uphole is towards the top of the upper photos, and towards the viewer in the lower photo.

Figure 2D.13 Plumes record rapid fracture propagation, but some disc fractures form incrementally, as shown by the ribs/arrest lines recording cyclic and circular fracture propagation just inside the outer core surface in these two examples. The upper photo is from a vertical core cut from limestone, the lower photo is of a horizontal core cut from a marley shale. In the numerous examples of ribbed, core-normal discs found in the horizontal core containing the example illustrated in the lower photo, the ribs typically originate at one or both sides of the core and grow upward, but occasionally they grow downward. The bottom half of this fracture surface is freshly broken rock. Similar ribbed discs are occasionally found on torque-related fracture surfaces described later, as well as on saddle-shaped disc surfaces that begin to resemble ribbed petal fractures. Both of these examples are from 4-inch diameter cores; up-section in the vertical core shown in the upper photo is towards the viewer; up-section in the horizontal core shown in the lower photo is towards the top of the photo and uphole is towards viewer.

Figure 2D.14 A ring of drilling mud invaded the disc fracture aperture near the core surface. This must have happened early in the history of the core while the drilling mud was liquid and before it dried, and suggests that the core across the disc fracture was incompletely separated, with the initial fracture existing only as a narrow crack in the rock. Vertical 4-inch diameter core; uphole is towards the viewer.

Figure 2D.15 Coned disc fractures, such as this one in a core cut from a calcareous marine shale, may be concave up or concave down. Ron Nelson (personal communication) has suggested several possible origins for coned fractures, including core barrel flexure.

Figure 2D.16 Coning may occur by itself or it may be superimposed on disc fractures as shown by these three views of a core cut from a shaley marine limestone. Three-dimensional cones such as those shown here may be related to slaking and core degradation by dehydration. Butts from vertical, 4-inch diameter core; uphole is towards the top of the left and upper right photo, and towards the viewer in the lower right photo.

Figure 2D.17 Some core-normal cracks in horizontal cores do not cut completely across the core, as shown by the pencil lines drawn adjacent to cracks in this core cut from a silty dolomite. These fractures probably form due to the high stresses at the bit–rock interface at the bottom of a wellbore during coring. Slabs from a horizontal, 3-inch diameter core; stratigraphic up is towards the top of the photo, uphole is to the left.

Figure 2D.18 Many disc fractures do not completely disarticulate the core, remaining as cracks in the intact rock that may be filled with drilling mud and/or a pseudo-mineralization of rock flour derived from the slabbing and sanding process. These can be mistaken for natural fractures unless the disc surfaces are examined. This example is in core cut from a marine shale; the core butt on the left was intentionally opened along the crack indicated by the arrow in order to examine the fracture surface, revealing the characteristic core-confined plume structure of an induced disc fracture. Butts of vertical, 4-inch diameter core; uphole is towards the top of the left photo, and towards the viewer in the right photo.

Figure 2D.19 Two views of a disc fracture; plumes that form readily on disc fractures in shale lithologies typically do not form in some lithologies such as the anhydrite shown in this photo pair, where it becomes difficult to distinguish between disc fracturing and handling breaks along bedding weakness planes. Slabs from vertical, 4-inch diameter core; uphole is towards the top of the left photo, and towards the viewer in the right photo. The arrow indicates the surface shown in plan view on the right.

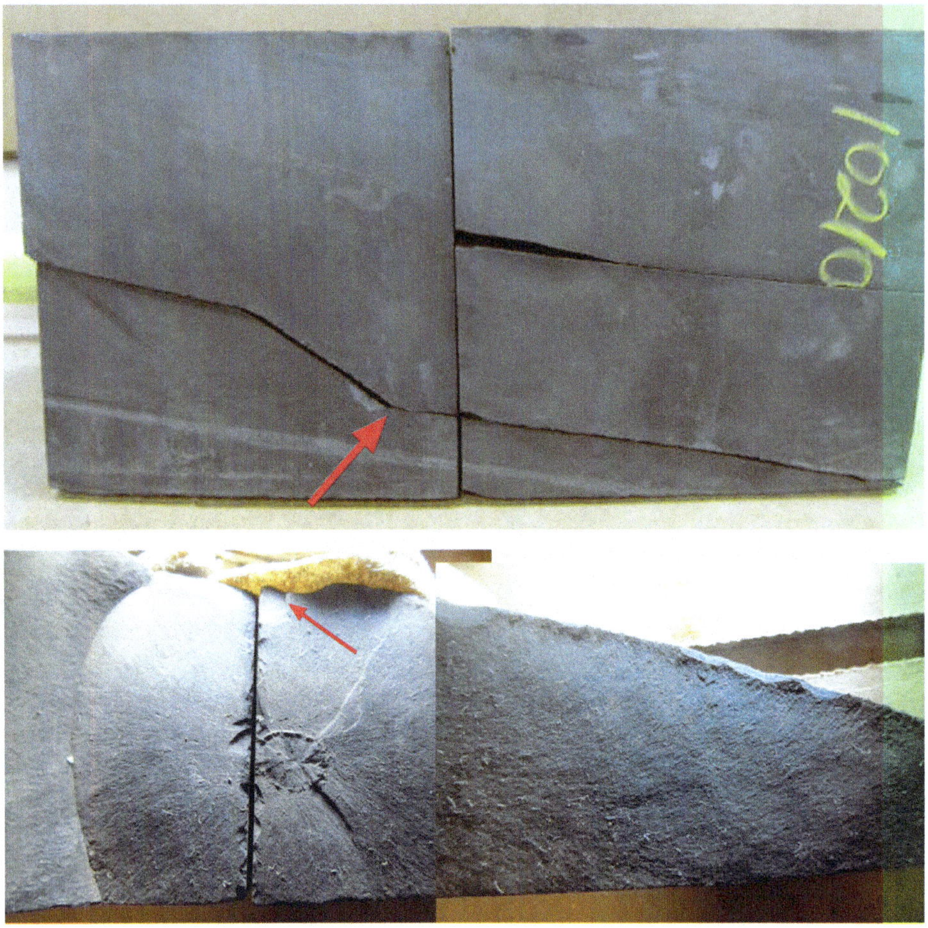

Figure 2D.20 Bed-parallel disc-type fractures can also form in horizontal cores, with similar characteristics and presumably due to the same mechanism (i.e., release from the weight of the overburden). This near-horizontal core from a marine shale cuts down-section to the left at a 10–12° angle to bedding (note the light bedding streak at the lower left of the upper photo). The kinked plane near the low side of the core in the upper photo is composed of two disc fractures joined by an inclined break in the core. The lower photo montage shows the origin of the lower disc fractures at a fossil and the plumed fracture surface, as well as the lip at the junction between the disc fracture and the core surface. The lack of similar interaction between the plume and the slab face indicates that discing occurred prior to slabbing. The arrow points out the limit of leftward fracture propagation. Slabs from a horizontal, 4-inch diameter core; stratigraphic up is towards the top of the upper photo and towards the viewer in the lower photo; uphole is to the right.

Figure 2D.21 A definitive bed-parallel disc fracture from a horizontal core, illustrating the plume that radiates to meet the core edge at a right angle and the interaction lip at the junction between the disc fracture and the core surface, both indicating that the disc formed after the core had been cut. Horizontal, 4-inch diameter core; stratigraphic up is away from the viewer, uphole is to the right.

Figure 2D.22 Modeling suggests that a significant stress concentration occurs at the interface between the core bit and the formation (e.g., Li and Schmitt, 1998), and that if the stresses are oriented properly and of sufficient magnitude, an induced fracture can form normal to the core axis regardless of bedding. This occurs in both vertical and horizontal core, as illustrated in this horizontal core cut from a siliceous shale. The natural fracture at the high-side of the core (top of the photo) is confined to a thin siliceous bed and cuts oblique to the core axis. The induced fracture, with a complex plume and rib pattern, strikes 30° to the natural fracture and normal to the core axis. Horizontal, 4-inch diameter core; up-section is towards the top of the photo, uphole is towards the viewer.

References

Kulander, B.R., Dean, S.L., and Ward, B.J., 1990, Fractured Core Analysis: Interpretation, Logging and Use of Natural and Induced Fractures in Core. AAPG Methods in Exploration Series 8. Tulsa: American Association of Petroleum Geologists.

Li, Y., and Schmitt, D.R., 1998, Drilling-induced core fractures and in situ stress. Journal of Geophysical Research, 103(B3), 5225–5239.

Obert, L., and Stephenson, D.E., 1965, Stress conditions under which core discing occurs. Transactions of S.M.E., 232, 227–235.

2E

Scribe-Knife Fractures

Scribe-knife fractures form in an oriented core when triangular steel scribe knives in the orientation shoe at the bottom of the core barrel cut too deeply into a core surface. This happens at the bottom of the hole as the core enters the core barrel. Ideally, the scribe knives should merely score the core surface, but one or more of the knives may wedge into the core with enough force to scar the core with a linear series of small spalls (Figure 2E.1), and they may even split the core, creating irregular fractures (Figures 2E.2–2E.4) (Lorenz et al., 1990).

Such fractures can resemble natural fractures, especially if viewed only where a slab face cuts the fracture plane, but they typically consist of irregular and unmineralized fresh-break surfaces that are easily distinguished from natural fractures if they can be traced back to an origin at a scribe line. It is important to recognize that as with all other types of induced fractures, rock flour from the slabbing process and drilling mud can be emplaced into the fracture creating a pseudo-mineralization.

Most scribe-line fractures propagate from the surface towards the center of the core (Figure 2E.5), and thus provide no useful information for core analysis. Where the in situ stress anisotropy is large, however, scribe-knife fractures may propagate parallel to the maximum horizontal compressive stress and can provide an orientation reference similar to but less reliable than that offered by petal fractures (Lorenz et al., 1990).

Many companies will not run oriented cores in some formations because the scribe knives damage the core as illustrated here, potentially creating core jamming and short core runs. Other companies routinely and successfully run scribe knives on all cores, whether oriented or not, providing an important orientation reference on the cores.

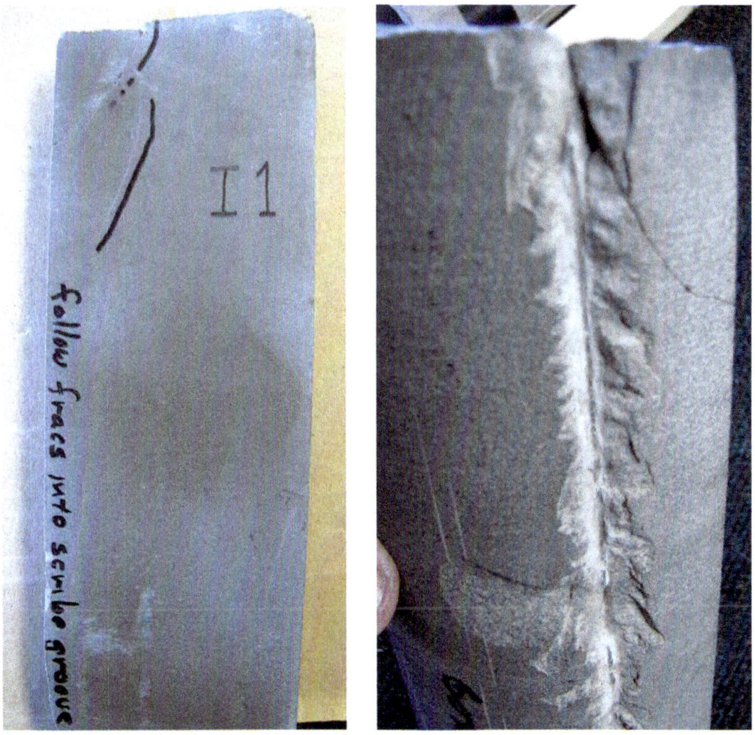

Figure 2E.1 Two views of a core piece containing scribe-knife fractures. Left: two irregular fractures adjacent to the black lines drawn on the core are exposed on the slab plane, and emanate from the badly scarred scribe-line groove on the back of the same slab (right) in this sandstone core. The scribe-knife fractures can be traced around the end of the core and back to the scribe groove. Slab from a vertical, 4-inch diameter core; uphole is towards the top of both photos.

Atlas of Natural and Induced Fractures in Core, First Edition. John C. Lorenz and Scott P. Cooper.

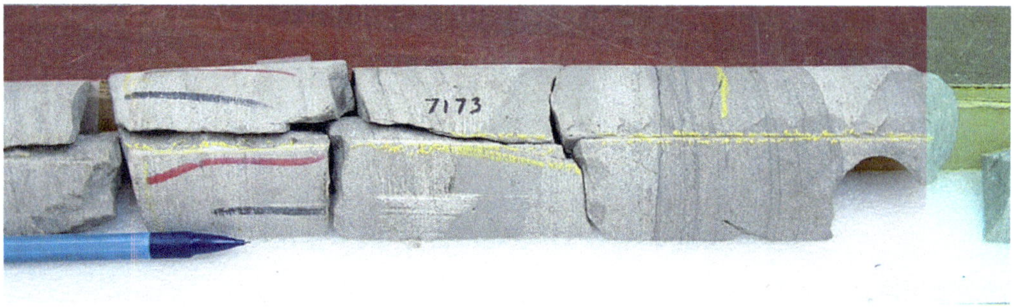

Figure 2E.2 An irregular scribe-knife fracture that can be traced for several feet along the length of a core cut from a well-cemented sandstone. The fracture originates at the scribe groove. Reassembled pieces from the butt of a 4-inch diameter vertical core; uphole is to the left.

Figure 2E.3 Three views of a scribe-knife fracture (red arrows) emanating from a scribe-line groove (black arrows) on the outer surface of a core cut from a limestone. Upper left: core with the fracture held closed, showing the scribe groove on the core surface; upper right: the core opened slightly along the fracture; bottom: the fracture face, showing irregularities along the scribe groove that become more planar as the fracture extends into the core where it was cut at a low oblique angle by the slab face. Vertical, 3-inch diameter core; uphole is away from the viewer in the top two photos, and towards the top of the lower photo.

Figure 2E.4 Two views of an irregular scribe-knife fracture (red arrows) originating from a damaged scribe groove (black arrow). The scribe-knife fracture propagated into the core normal to the core surface and intersected/terminated at a natural fracture (yellow arrows)

Figure 2E.5 Two scribe-knife fractures that propagate from scribe lines on different sides of a small-diameter well-cemented sandstone core and that meet and join near the center of the core. Vertical, 2.5-inch diameter core; uphole is away from the viewer.

Reference

Lorenz, J.C., Finley, S.J., and Warpinski, N.R., 1990, Significance of coring-induced fractures in Mesaverde core, northwestern Colorado. American Association of Petroleum Geologists Bulletin, 74, 1017–1029.

2F

Torque and Helical Twist Fractures

When a drill string is rotated from a kelly table or similar drive at the surface, the bit and outer core barrel rotate "to the right" (i.e., clockwise looking downhole) at about 60 revolutions per minute. Bearings in the tool separate the rotating core bit and outer barrel from the stationary inner core barrel and, if used, from the core-scribing shoe. The core, which is not yet detached from the formation below the bit, does not rotate with the drill string as it is cut and enters the core barrel.

The system is very good but not perfect, and the supposedly stationary parts of the coring tool are dragged clockwise by friction in the system, usually slowly, a couple of degrees of rotation per foot of core, but if the bearings in the core tool wear out or get contaminated, the inner barrel can rotate more rapidly, imparting torque to core in the barrel and setting up rotational shear stresses between the core in the barrel and core that is still attached to the formation. This twists the core around its longitudinal axis. This rotation is recorded by the varying degrees of spiral in the scribe grooves on an oriented core, but it also occurs while cutting unoriented core.

Spinoffs form along a core when twisting stresses are relieved by the formation of core-normal rotational shear planes, as described later in Part 3. Torque structures can also form where torque on the core was more subtle. Breaks can form between the stationary and rotating parts of the core where the two core sections do not spin continuously against each other, the structures typically consisting of helical fracture planes that spiral about the core axis, (Figures 2F.1–2F.3). The plume structure and lips that mark some of the shallow-dip-angle torque fractures (Figure 2F.4) resemble the fractographic markings on many disc fractures, suggesting a genetic relationship and that some types of disc and torque fractures are part of a spectrum of induced fractures.

Other torque fractures dip quite steeply (Figures 2F.5, 2F.6), forming isolated helical planes in a core. Similar helical fracture planes can be formed by twisting a piece of chalk (the hard part of this experiment is holding the two ends of the chalk tightly enough to impart the surprising amount of torque required to break it).

If there is significant weight on the torqued zone of the core, due perhaps to a tall column of core in the barrel above it or sometimes due to jamming of core in the barrel, shear zones composed of numerous helical fracture planes can form (Figures 2F.7, 2F.8). One of the more spectacular examples was the product of inadvertently trying to force 32 feet of core into a 30 foot barrel when the driller lost track of his depths. As a result, when recovered, the bottom 4 feet of the core were shattered in a very systematic pattern of helical shards that resembled a pine cone when removed from the barrel, and that quickly fell apart (Figure 2F.9).

Atlas of Natural and Induced Fractures in Core, First Edition. John C. Lorenz and Scott P. Cooper.

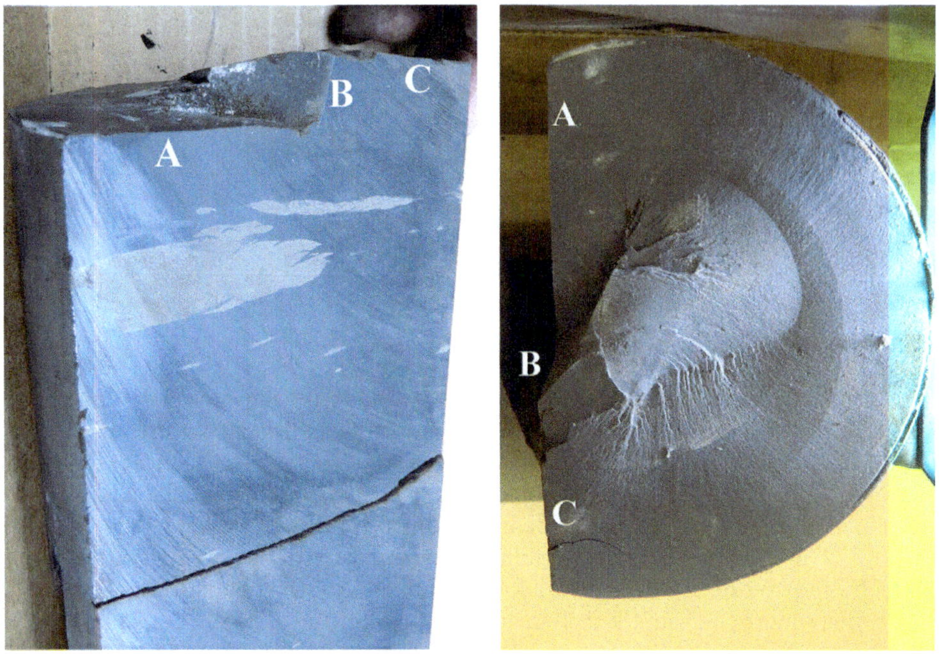

Figure 2F.1 Two views of a low-angle inclined helical torque fracture in core from a marine shale, viewed on the slab face of the butt (left) and end-on in the same piece of core (right). Twisting broke the core but did not continue to spin the top core piece against the bottom piece, which would have formed a spinoff plane marked by circular patterns on a core-normal shear surface. The helical plane created by twisting climbs up-section at a low angle from A around the core circumference to C. The core later broke along the inclined plane B to connect the high and low sides of the helix. Note the lip at the core edge, indicating the torque fracture formed after the core was cut. Butts from vertical, 4-inch diameter core; uphole is towards the top of the left photo and towards the viewer in the right photo.

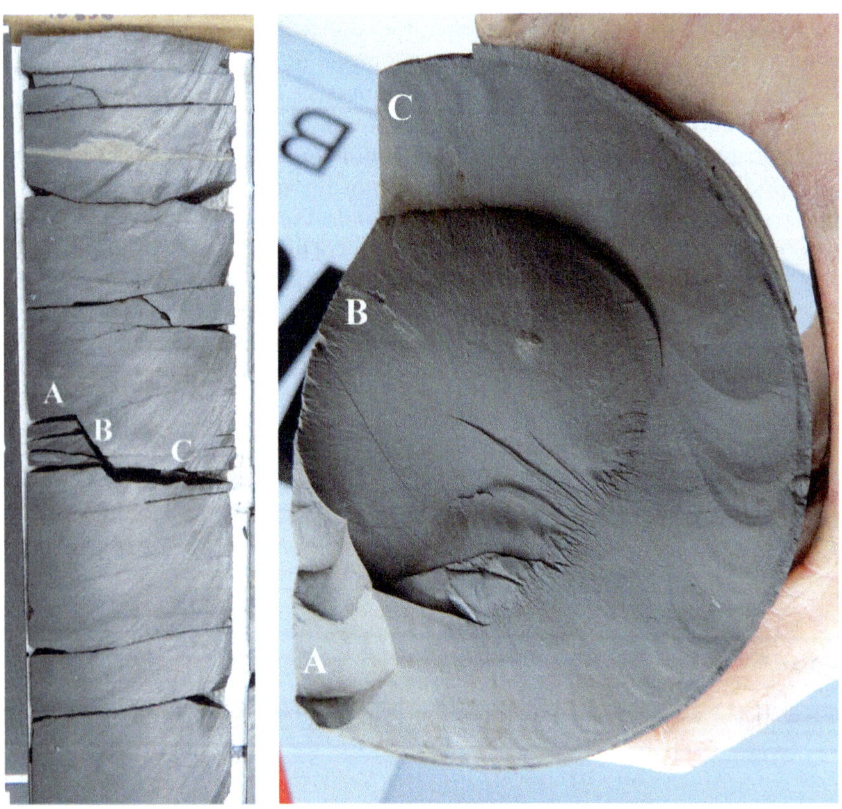

Figure 2F.2 Two views of a spiraling helical torque-induced fracture plane in core cut from a muddy shale. Left photograph shows the expression of this circular, inclined-plane fracture on the two-dimensional slab face, cutting that surface at levels A and C as it spirals around the core. The helical plane dips at a shallow angle, and the two sides are joined across the core by the inclined break in the rock at plane B. The right-hand photo is this same fracture, viewed from the top face-on in the equivalent butt section of the core. Note the subtle ribs on the fracture plane at the upper right side of the right-hand photo; ribs indicate incremental fracture propagation around the edge of the core and formation of the helix after the core was cut. They are very similar to the ribs found on some of the disc fractures described earlier, suggesting some continuity in the spectrum of fracture forms between disc and torque fractures. Vertical, 4-inch diameter core; uphole is towards the top of the left photo and towards the viewer in the right photo.

Figure 2F.3 A less definitive torque fracture plane that is nearly horizontal and normal to the axis of core cut from a shale, but that shows similar fractography including a ribbed lip at the margin of the core surface. Vertical, 3-inch diameter core; uphole is towards the top of the photo.

Figure 2F.4 A torque fracture in a silty shale, marked with a plume showing propagation of the fracture within the core, and a lip at the edge of the core showing that the fracture propagated after the core had been cut. The shadow in the lower right corner of the photo shows the relief on the fracture face as it climbs, spiraling clockwise uphole. Vertical, 4-inch diameter core; uphole is towards the viewer.

Figure 2F.5 Some torque fractures such as this one, formed in core cut from a muddy marine shale, consist of single, isolated helical planes with much steeper dip angles. Vertical, 3-inch diameter core; uphole is towards the top of the photo.

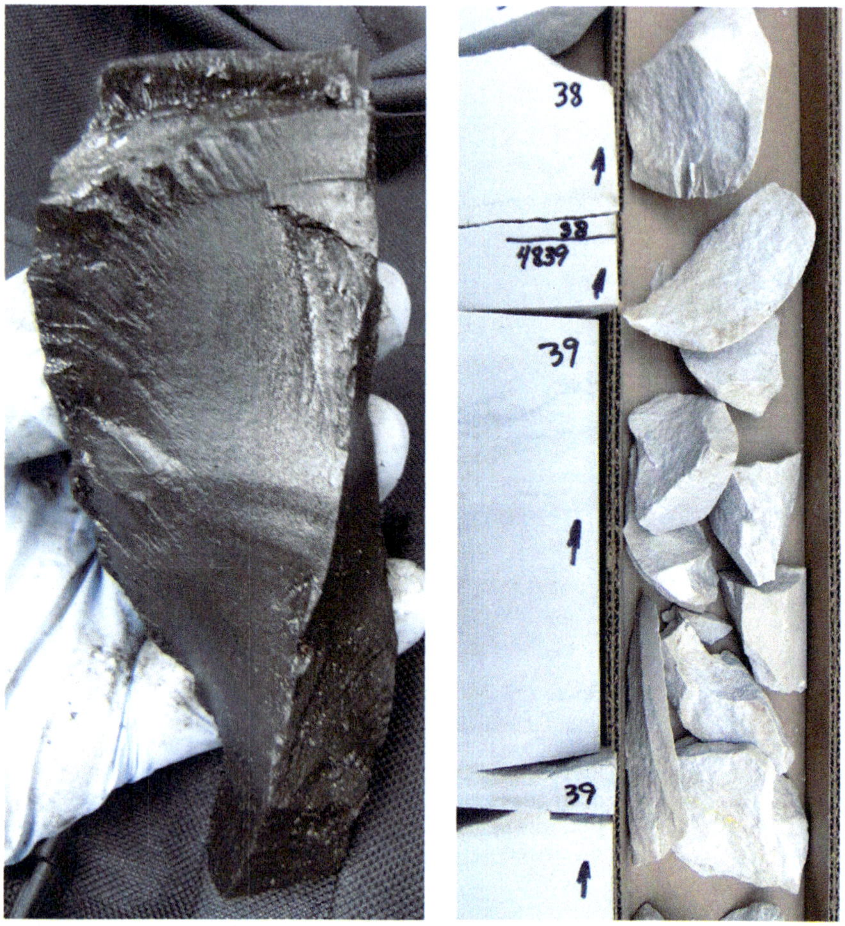

Figure 2F.6 Core shattered by torque fractures consists of shards that are not held together by anything except dried drilling mud and wrappings of plastic film or aluminum foil. Once slabbed, the overall pattern may be lost although the individual shards still have a characteristic helical shape. Left a large helical shard from a marine shale; right: remnant helical shards from a torque-shattered zone in a sandstone core. In the example on the right, the service company had no intact core to slab and had no choice but to place representative helical shards in the box in lieu of slabs. Shards from vertical, 4-inch diameter cores; uphole directions on the individual shard are unknown.

Figure 2F.7 Two views of helices on torque fractures that remained relatively intact during slabbing of this anhydrite. Shards may spiral both clockwise and counter-clockwise in the same core interval, producing complex intersecting fracture patterns. The right photo illustrates the steeply dipping helical fracture planes that are not obvious on the slab face. Butts from a vertical, 4-inch diameter core; uphole is towards the top of both photos.

Figure 2F.8 A similar pinecone texture formed by intersecting, counter-spiraling helical fractures in core cut from an anhydrite. The box on the left holds the shattered remnants of the core from the slab box. The butts of the core, on the right, held together during slabbing. Vertical, 4-inch diameter core; uphole is towards the top of the photo.

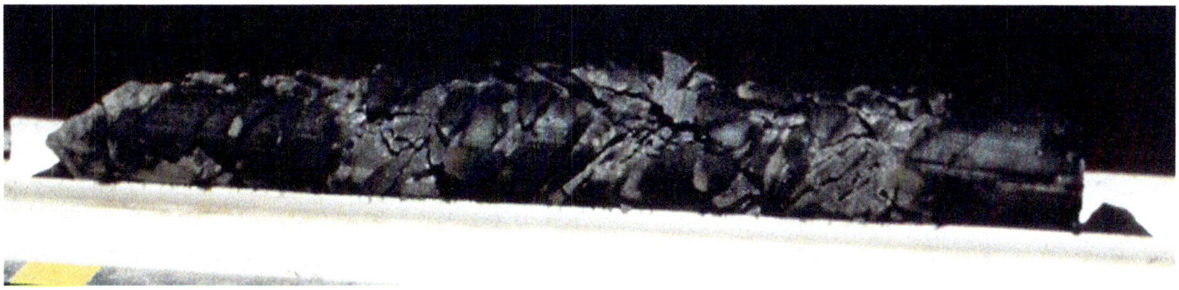

Figure 2F.9 A 2-ft interval of siltstone at the base of this core was shattered by twisting when the barrel was overfilled due to a miscalculation of depths. This interval consists of diagnostic helical shards of core. Vertical, 4-inch diameter core; uphole is to the left.

2G

Core-Compression Fractures

Even without the complication of torque, induced fractures can be created by core-parallel compression when a core barrel is overfilled (Figures 2G.1, 2G.2). This can occur when coring continues beyond a planned 30-ft interval, less likely with modern drill rigs but which can still found in archived core.

Core-parallel compression fractures can also be created when the basal part of one core is inadvertently left in the hole, unknown to the rig personnel, and partially fills the next 30-ft core barrel before beginning to cut the next core interval. Be cautious when a core run comes up short as the next run may come out long and include induced compression fractures. Core retrieval and processing companies have been known to move the excess core from an overly long core run to the next short run, equalizing the core lengths in the two runs so that although the on-site core runs came up with 29 feet and 31 feet of core, they may be marked during processing as two 30-ft runs, leaving the analyst in the dark about the origin of a zone of compression fractures between the two runs. It is best to be on site when cutting core to make sure that all is going as planned during cutting and handling of your core, or to recognize when it doesn't and understand the ramifications.

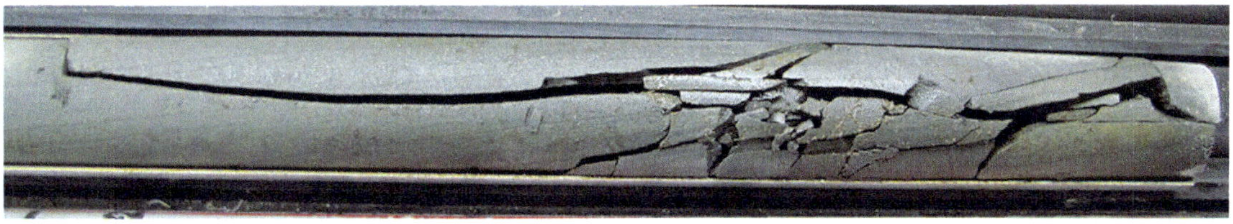

Figure 2G.1 A fracture zone in whole, horizontal core from an epimarine siltstone, as displayed immediately after retrieval from the wellbore and logged at the rig site. The core pieces are preserved in place by one half of the pre-split, aluminum core barrel liner. Compression parallel to the core axis due to an attempt to put more core into the core barrel than the barrel length would accommodate caused the core to split along and subparallel to bedding. The secondary shears cutting oblique to bedding form compression-related wedges in the core. Attempts to retrieve the core by pushing it out one end of an unsplit inner barrel would have resulted in core rubble. This zone is shown in the slab box in the next figure; note that the light-yellow low-viscosity oil that coats the whole core in this on-site photo is missing after slabbing. When possible, try to view a core before slabbing. Horizontal, 2.5-inch diameter core; stratigraphic up is towards the top of the photo, uphole is to the left.

Figure 2G.2 The same core interval shown in the previous photo, but after slabbing. The major features are still present, but all the smaller pieces of the puzzle have been lost. Horizontal 2.5-inch diameter core, cut from a siltstone. Stratigraphic up is towards the top of the photo, uphole is to the left.

Atlas of Natural and Induced Fractures in Core, First Edition. John C. Lorenz and Scott P. Cooper.

2H

Percussion-Induced Fractures

Hammers of various sizes ranging from big to bigger are indispensable tools on a drilling rig floor and in many stages of core processing. Hammers have been used for disassembling the core barrel and for breaking cores into proper lengths to fit into core boxes. Cores even used to be removed from the core barrel on site by hanging the barrel vertically from the rig draw-works so that the core could cascade out the bottom, to be caught at intervals with a clamp while the geologist picked up and boxed the pieces, mindful of the up direction for each core piece and trying to retain all 10 fingers. When the core hung up in the barrel, it was jarred loose with sledge-hammer blows to the outside of the barrel, and the loosened core then fell suddenly down on top of itself, producing percussion fractures as the core ends banged against one another.

Hammers are used less frequently with the advent of split aluminum core barrel liners and masonry chop saws, but there are still numerous older cores in the warehouses that were processed primarily with hammers that created numerous percussion-related induced fractures. Moreover, many cores are still removed from short sections of unsplit inner core barrels with metal rods inserted with vigor longitudinally into the barrels to push out the core, a process that also produces percussion fractures.

There are numerous opportunities for pieces of a core, especially horizontal cores, to come into percussive contact with each other during the coring process. Pieces of a horizontal core, separated from the formation by a vertical natural fracture, can slide up and down the barrel a few inches so that the ends on either side of the fracture knock abruptly against each other, producing percussion fractures.

Most percussion fractures are easily recognized by their poorly planar and sometimes conchoidal surfaces, and by origination points marked with crushed rock (Figures 2H.1–2H.3). The fracture surfaces commonly have lips where they intersect the outer core surface (Figure 2H.4). If the fracture is plumed, the plume axis commonly follows the core axis. Some percussion fractures form a spiraling plane along the core, suggesting the core was being twisted at the time of fracturing (Figures 2H.5, 2H.6), and suggesting a relationship to the previously described torque fractures. Repeated percussion on the end of a core may remove a significant volume of the core (Figure 2H.7). Since they are not directly related to the coring process, percussion fractures may propagate up-core, down-core, oblique to the core, or across the core.

Atlas of Natural and Induced Fractures in Core, First Edition. John C. Lorenz and Scott P. Cooper.

Figure 2H.1 Two views of an inclined, less-than-planar percussion fracture in a dense limestone. It is held slightly opened in the top photo, and positioned opened in butterfly fashion in the bottom photo to show the two fracture surfaces. The fracture originates at a crushed point of impact (arrows). The core is ringed by a shallow saw cut made at the rig site and intended to cut the core and core barrel liner into manageable lengths for transportation to the lab where the core would be extruded from the liner. However, the competent core did not break along the saw cut as intended and had to be broken with a hammer, producing this irregular percussion fracture. The scribe grooves along the core surface parallel to the core axis were made by scribe knives in a core orientation shoe. Vertical, 4-inch diameter core; uphole is to the left in the upper photo and towards the top of the left-hand core piece in the lower photo.

Figure 2H.2 Left: a whitish crush zone (arrow) marks the point of impact for a percussion fracture in core cut from a non-marine limestone. The fracture propagated down the axis of the core, with twist hackle forming where it intersected the free core surface. Right: a roughly planar fracture in a marine limestone, marked by plume structure and lips at the core surface, cannot be traced back to a similar point of origin since that piece of core is missing. Vertical cores; uphole is towards the top of both photos. Left photo shows 4-inch diameter core, right photo shows 3-inch diameter core.

Figure 2H.3 Two views of a core piece cut from a well-cemented sandstone and containing two percussion fracture planes that strike oblique to each other. Both appear to originate at the same crushed point of impact (arrows) although the first point of origin may have been lost after the second fracture formed. The fractures propagated down the core axis without forming a plume, and terminate blindly in the rock. Vertical, 4-inch diameter core; uphole is towards the top of the left photo, and towards the viewer in the right photo.

Figure 2H.4 A core-normal break across this sandstone core provided an opportunity for the two pieces of core to separate and slam back together, creating percussion-related spalls on both sides of the break. The spalls originated at the high point in the break near the red-black uphole orientation lines, with plume structure radiating both uphole and downhole from the point of impact. Vertical, 4-inch diameter core; uphole is to the left.

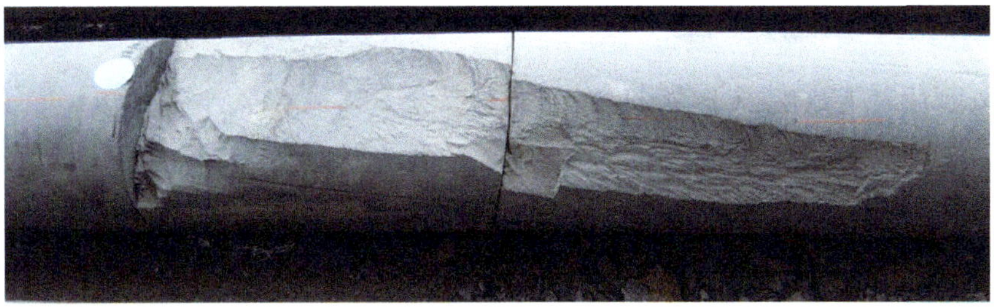

Figure 2H.5 This percussion fracture in a horizontal core, cut from a marine limestone, originated at a natural fracture (the cross-core plane at the left of the photo) when the core separated across the fracture and then was slammed back together. The induced fracture propagated along the core axis in one direction only. The fracture is slightly helical, suggesting that the core was being twisted at the time of fracturing. Since the natural fracture is oblique to the core axes, separation of the core along the natural fracture and slight rotation of the core pieces relative to each in the barrel during coring put non-mating surfaces in contact and provided the local point of impact for fracture origination. Some of the ribs have fracture surface has been washed out by erosion as drilling mud flowed along the fracture. Horizontal, 4-inch diameter core; uphole is to the left.

Figure 2H.6 Two views of conchoidal percussion fractures formed at a rotated core break (spinoff) in a horizontal core cut from limestone. The concentric circle patterns on the ends of the core indicate that the core pieces rotated against each other, but the core still locks reasonably well across the break, indicating that the amount of rotation was minimal. The asperities across the break impacted each other during rotation, spalling flakes off the edge of the core. Horizontal, 4-inch diameter core; uphole is towards the left in the upper photo, and towards the viewer in the lower photo.

Figure 2H.7 Repeated percussion fracturing in this horizontal core cut from a sandstone spalled off numerous pieces of the core below/to the right of the planar natural fracture, leaving a pointed, non-matching core end. Horizontal, 4-inch diameter core; uphole is to the left.

21

Bending Fractures with Barbs

A bending fracture (see Kulander et al., 1990) starts as a planar, core-normal tensile fracture but turns abruptly to form a single, distinctive tang or barb oriented at nearly right angles to the cross-core fracture plane. These fractures form as classic beam-bending fractures, developing when the core is bent to some small degree such as when a core barrel is pulled out of a wellbore around a small-radius heel in a deviated hole, when a bouncing core barrel is carried across a rig site balanced in the middle on the tines of a forklift and bouncing/flexing as if it's trying to fly, or sometimes when a core is bent and intentionally broken so that it will fit into a core box. One sandstone core we worked, cut in 1948, was segmented by bending fractures regularly spaced every 3 feet, probably bent and broken over the tailgate of a pickup truck in order to fit it into the core boxes.

Bending fractures form when the outer radius of a bent core is extended and fails as a true tension fracture normal to the core axis. Failure propagates across three-quarters of the core diameter but then extends into a zone of compression on the inside of the bend. Once in the zone of compression, the developing fracture curves to try to parallel the local compressive stress, propagating subparallel to the core axis and thus forming the distinctive barb (Figures 21.1, 21.2). The last bit of the fracture may make another abrupt turn to intersect the core surface at nearly right angles (Figures 21.3, 21.4). Similar fractures can be created by bending sticks of chalk.

These fractures are found in horizontal, deviated, and vertical cores, and are the product only of the local bending stresses; they are not related to regional or coring stresses. The barbs have no relation to bedding or to vertical, and can form pointing uphole or downhole, and parallel or oblique to bedding in a horizontal core.

The slab of a core may or may not cut across the barbs of these fractures, and without the barb it can be difficult to distinguish bending fractures from disc or other fracture types. Conversely, remnants of bedding in a horizontal core, petal fractures, and other structures (Figure 21.5) can form false bending barbs.

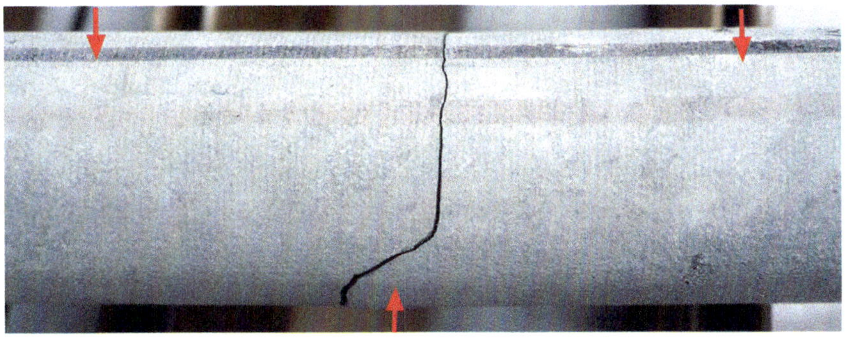

Figure 21.1 A bending fracture in horizontal core cut from a marine siltstone shows the classic profile. The core was bent enough to break the rock during processing, with the outer ends forced downward relative to the center as shown by the arrows. A planar, core-normal, tensile fracture propagated three-quarters of the way across the core diameter in the tensile regime on the outside of the bend, then entered the zone of compression on the inside of the bend and turned abruptly by almost 90° to follow the compressive stress trajectory. The fracture turned again in order to intersect the free core surface at a right angle. Horizontal, 2.5-inch diameter core; uphole is to the right.

Atlas of Natural and Induced Fractures in Core, First Edition. John C. Lorenz and Scott P. Cooper.

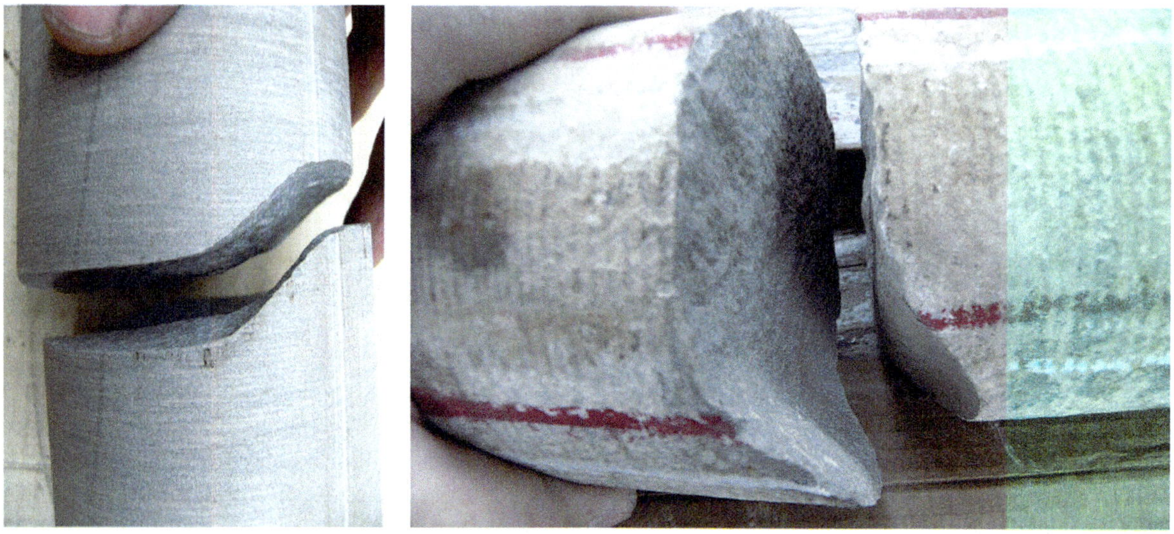

Figure 21.2 Bending fractures in vertical core cut from an eolian sandstone (left) and horizontal core cut from a marine siltstone (right) both show the characteristic profile of this type of induced fracture. Both cores are 4 inches in diameter, uphole is towards the top of the left photo, and towards the left of the right photo.

Figure 21.3 Some bending fractures have shapes that are variations on the classic form, including irregularities such as this bending fracture from a well-cemented sandstone. Vertical, 4-inch diameter core; uphole is towards the top of the photo.

Figure 21.4 A three-dimensional view of a bending fracture in a well-cemented eolian sandstone shows the mud-stained planar tensile surface, the abrupt 90° bend as it enters the compressive zone of the bend, and a second, smaller abrupt bend where it cuts normal to the core surface in order to complete the break in the core. Vertical, 4-inch diameter core; uphole is to the upper right of the photo.

Figure 21.5 Other structures can resemble bending fractures with barbs, and the core must be examined carefully. The barbed structure in the vertical limestone core on the left consists of a petal fracture intersected by a core-normal saw cut. The barbed structure (red arrow) in the sandstone core on the right was formed by the intersection of one of a series of inclined, calcite-mineralized natural fractures (black arrows) and a handling-related induced fracture cutting across the core axis. Four-inch diameter cores; uphole is towards the top of both photos

Reference

Kulander, B.R., Dean, S.L., and Ward, B.J., 1990, Fractured Core Analysis: Interpretation, Logging and Use of Natural and Induced Fractures in Core. AAPG Methods in Exploration Series 8. Tulsa: American Association of Petroleum Geologists.

2J

Irregular Crack Networks

Cores cut from dense, fine-grained limestones and dolomites often host poorly defined, irregular systems of intersecting to subparallel induced cracks of uncertain origin. Some of the cracks resemble petal and disc fractures, and locally they form arrays pointing inward towards an unfractured central zone in the core (Figure 2J.1). Other crack sets form seemingly random networks (Figure 2J.2).

The individual cracks in a system may be single planes or may be multi-stranded. The rock parts easily along the cracks but for the most part the core remains intact (Figures 2J.3, 2J.4). Where stylolites are present, the cracks may follow the stylolites (Figure 2J.5), cutting across the stylolite teeth. Stylolite planes may offer a mechanical weakness plane that can be exploited, or they may concentrate the crack-forming stresses. The crack apertures often appear to be mineralized due to the presence of rock flour from slabbing and sanding, and/or due to damage to the rock immediately adjacent to the fracture caused by these processes.

The resemblance of some of these cracks to petal fractures (see Figure 2J.4) and disc fractures (Figures 2J.6, 2J.7), their lack of mineralization, and local patterns that interact with core surfaces all indicate that the crack networks are induced and that they are part of a spectrum of fractures rather than discrete fracture types. Most of these fractures are probably related to stress concentrations at the core–bit/rock interface.

Figure 2J.1 Radial arrays of cracks in limestone (left) and dolomite (right) that extend inward towards a less fractured central zone. Both examples are from the slabs of vertical, 4-inch diameter core; uphole is towards the top of both photos.

Atlas of Natural and Induced Fractures in Core, First Edition. John C. Lorenz and Scott P. Cooper.

Figure 2J.2 A highly irregular system of cracks, some following the stylolite planes, in a core cut from a limestone. The core has been wetted and the water has partially evaporated, with the cracks retaining water longer than the rest of the slab surface. Slab from a 4-inch diameter vertical core; uphole is towards the top of the photo.

Figure 2J.3 The cracks are typically also irregular in the horizontal plane, as shown by this cut across a limestone core. Vertical, 4-inch diameter core; uphole is towards the viewer.

Figure 2J.4 Crack networks may resemble induced petal and saddle fractures, as do these cracks in a limestone. The cracks initially angle inward and downward from the edge of the core, but the dip angles diminish and the cracks flatten as they cross the middle of the core. Slab from a vertical, 4-inch diameter core; uphole is towards the top of the photo.

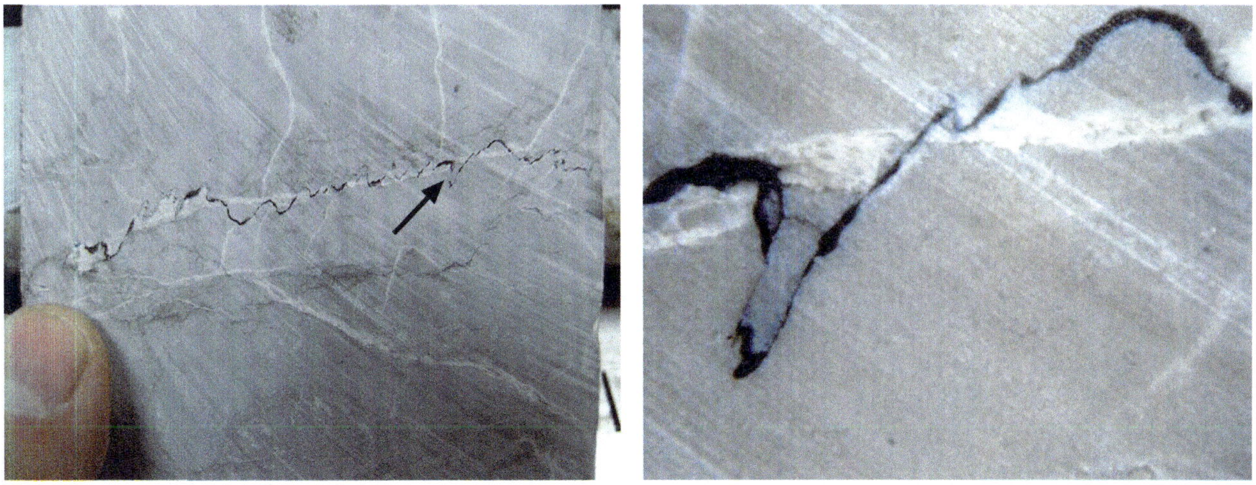

Figure 2J.5 Irregular cracking in a limestone core follows a weakness plane provided by a stylolite, cutting across the stylolite teeth. Black arrow in the left photograph shows the location of the enlarged photograph on right. Slab from a vertical, 4-inch diameter core; uphole is towards the top of the photo.

Figure 2J.6 A systematic set of core-normal cracks (arrows) in a fine-grained dolomite resemble incomplete disc fractures. The unmineralized surface of one fracture (red arrow on the left photo) is shown in the photo on the right, and displays no particular diagnostic shape or fractography. Vertical, 4-inch diameter core; uphole is towards the top of the left photo, and towards the viewer in the right photo.

Figure 2J.7 A system of irregular tan-colored cracks cuts across a dolomitic limestone core. The wellbore is vertical but bedding is inclined, dipping to the left, and the calcite-mineralized bed-normal natural fractures dip to the right. The calcite mineralization is a brighter white where the induced cracks cut across the natural fractures due to local small-scale shattering of the calcite. Slab from a vertical, 4-inch diameter core; uphole is towards the top of the photo.

2K

Induced Fractures with Curved Strikes

Other induced fractures, with curved strikes and of uncertain origin, occur in a few cores. One type, noted in only one of many cores logged over the years but occurring repetitively in that core, has faint concave-upward ribs marking the fracture face, closely resembling a centerline fracture that both originated and terminated blindly in a core (Figures 2K.1, 2K.2). The reason for the strike curvature is unclear but these are probably a rare form of centerline fracturing. Core continuity was insufficient to determine whether or not the five logged curved fractures in this core had similar orientations as would be expected of centerline fractures, but all five examples were concave towards the nearest outer surface of the core (see Figure 2K.2). These fractures may have formed during coring by lateral extension of the unconfined rock in the core barrel under the weight of the overlying core sections, and they may be related to the structures illustrated in Figures 2C.19, 2C.20, and 2C.21.

Another type of curved-strike induced fracture occurs in cores from several different formations, and is concave towards the middle of the core (Figures 2K.3, 2K.4). The fractures form curved chips or flakes of core half a centimeter to a centimeter thick and resemble the elongated curved plane of a posthole shovel blade. One side of the flake is formed by the outer surface of the core, the other side consists of the curved plane of fracturing and is nearly parallel to the outer core surface. The fracture planes extend around a quarter of the circumference of the core, and extend along the core axis for a height of several inches to several feet. Lips form where the fractures intersect the core surface.

Oilfield rumor relates these fractures to core spalling caused by degassing of the core, but there is no evidence to support that mechanism. The fractures may be related to stress release, to thermal stresses due to changing temperature in the rock as it comes to the surface, or perhaps to dehydration; evidence such as a consistent or random orientation relative to the in situ stresses would help in the development of a theory of origin for these structures. In several examples the curved induced planes appear to extend from, or at least to be associated with, very narrow, mineralized fractures located near the edge of the core and parallel to the overall strike of the induced planes (see Figure 2K.4).

Atlas of Natural and Induced Fractures in Core, First Edition. John C. Lorenz and Scott P. Cooper.

Figure 2K.1 Two views of one of several induced fractures with blind top and bottom terminations and slightly curved strikes in a core cut from a marine shale. The fracture changes strike by about 15° within this section of the core. Butts from a vertical, 4-inch diameter core; uphole is towards the top of both the photos.

Figure 2K.2 Top: end-on view of another curved fracture (white arrow) found in the same shale core, showing a 30° change in strike. The bottom photo shows the face of the fracture, with concave-upward ribs (visible only under oblique lighting) and a edge effect where it intersects the outer core surface on the right side of the photo. Vertical, 4-inch diameter core; uphole is towards the top of the photo.

Figure 2K.3 Two views of an induced fracture with a curved plane that parallels the outer core surface, in a core cut from a muddy marine shale. Vertical, 3-inch diameter core; uphole is towards the viewer in the upper photo, and to the right in the lower photo.

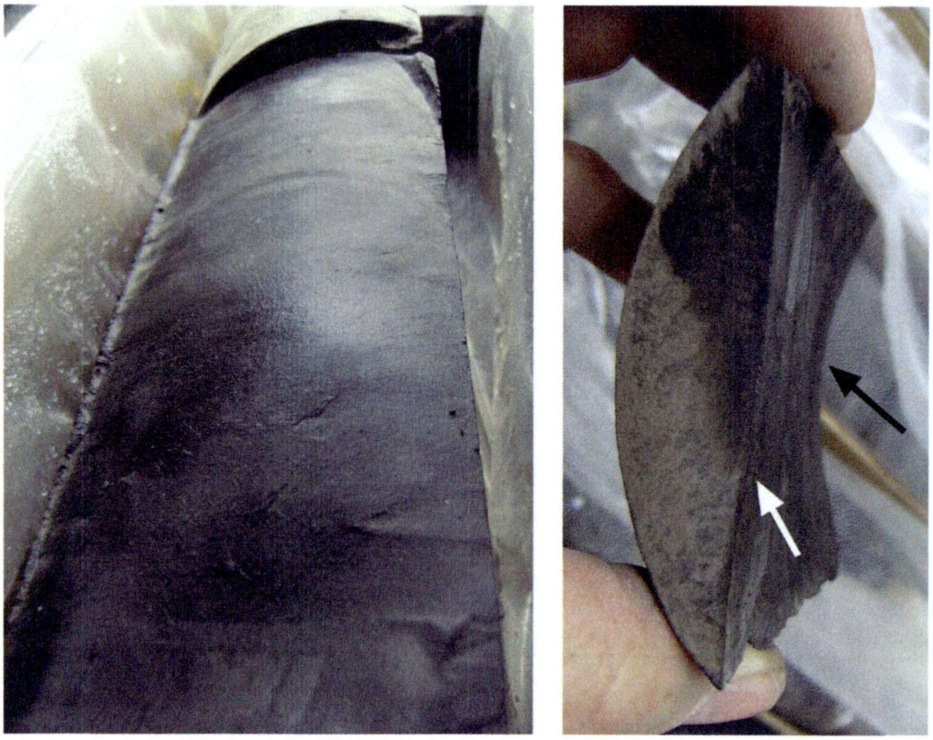

Figure 2K.4 Left: a view along the core axis of a tall induced fracture, showing an unornamented face and the lip or cusp that formed at the intersection between the fracture and the core surface, especially on the left side of the photo. Right: edge-on view of the same fracture further along the core, showing the planar, lightly mineralized natural fracture (white arrow) extending downward into the curved induced fracture plane (black arrow). Vertical, 4-inch diameter core; uphole is away from the viewer in the left photo, and toward the viewer in the right photo.

2L

Waterflood-Related Fractures

A unique system of short, parallel, unmineralized fractures (Figures 2L.1, 2L.2) was captured in cores cut from a sandstone oil reservoir. This system has characteristics that are consistent with an origin as a fracture dilatancy fabric created by the long-term, high-pressure, high-volume waterflood program that was carried out in this field. The fractures are pervasive in the 600 ft of cores cut from the poorly cemented, massive, fine-grained sandstones after the waterflooding program had ended. During the waterflood, injection pressures were high but did not exceed the local parting pressure, and these are not hydraulic fractures. Rather, flooding raised the formation pore pressure, counteracting a percentage of each of the three confining stresses.

Rock becomes brittle and susceptible to extension fracturing under these conditions (see Lorenz et al., 1991; Rhett, 2001; Robinson, 1959). The induced fractures in the cores, with characteristics that are obviously different from the conjugate natural fractures that are also present in the cores, are irregularly planar and consist of zones of vertical fracturing, zones of horizontal fracturing, with transition zones of inclined fractures, suggesting that they formed under conditions of low effective confining stress with the three confining stresses being nearly equal, also consistent with the waterflood history.

The induced fractures are not mineralized, the rock is still partially intact across these fracture planes, and the well-developed fractures create a dilatancy texture of pervasive cracking. The fracture planes are often connected in an anastomosing pattern of low-angle intersections. Individual fracture planes are rough and irregular, range from less than an inch to several feet in height, and have parallel to subparallel strikes.

The horizontal induced fractures are similar, following or being subparallel to bedding (Figure 2L.3). These fractures are commonly short and many do not cut entirely across the core. As many as 32 horizontal induced fractures can occur in a single foot of core length although a 1-inch spacing is more common. Horizontal and vertical fractures do not occur in the same intervals of core. A few zones of inclined fractures are present, possibly due to the influence of the associated inclined bedding, but also formed as transitional areas between horizontal and vertical fractures.

The induced fractures show obvious interactions with the natural fracture planes, indicating that the natural fractures were present prior to propagation of the induced fractures (Figure 2L.4).

Atlas of Natural and Induced Fractures in Core, First Edition. John C. Lorenz and Scott P. Cooper.

Figure 2L.1 Irregularly planar, parallel strikes of vertical fractures shown in the cross-section of a core cut from poorly cemented sandstone. Note the tough layer of obscuring salty efflorescence that could not be brushed away at the left edge of the core. The brown line across the right side of the photo is rust from the iron channel of the core tray, left on the core when it was set upright on the channel while wet. Butt from a vertical, 4-inch diameter core; uphole is towards the viewer.

Figure 2L.2 Vertical, dilatancy extension cracks of the vertical fracture system in the poorly cemented sandstone. Slab of a vertical, 4-inch diameter core; uphole is towards the top of the photo.

Figure 2L.3 Irregular horizontal dilatancy cracks (left) and inclined cracks (right) in a poorly cemented sandstone. Both overlie an unfractured, clay-flecked/laminated sandy lithology. The inclined fractures appear to have been diverted to follow inclined, possibly deformed bedding planes. Slabs of a vertical, 4-inch diameter core; uphole is towards the top of both photos.

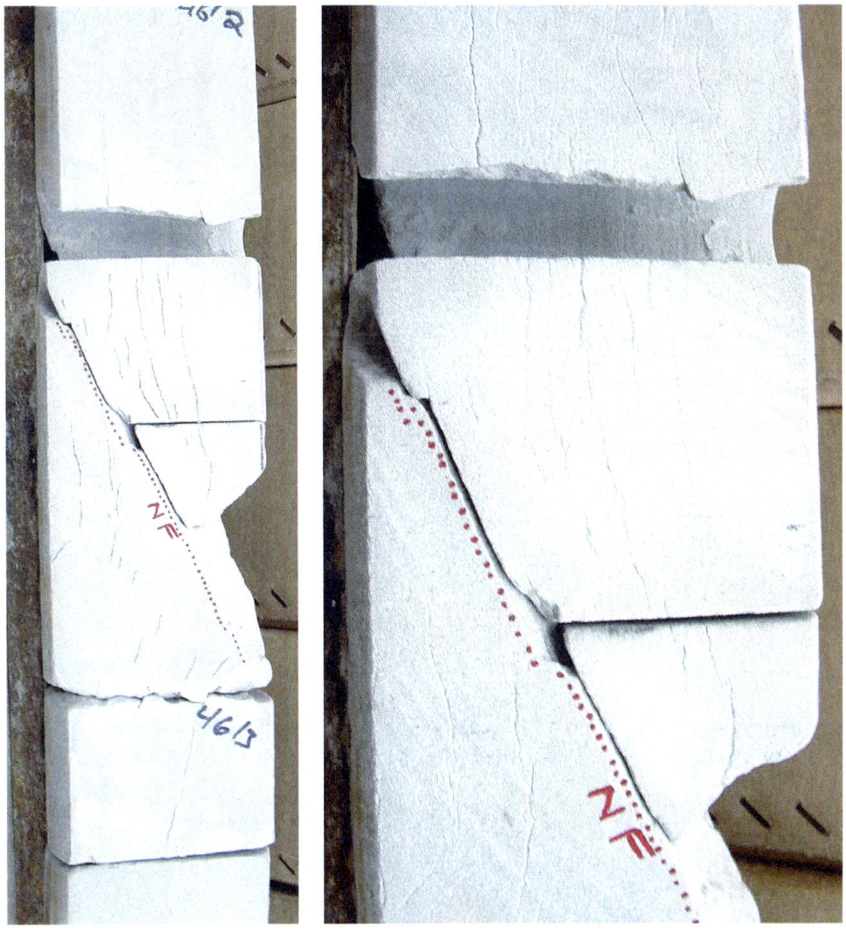

Figure 2L.4 The waterflooding-related induced fractures, highlighted in pencil on the left and shown in close-up on the right, curved as they interacted with a natural fracture ("NF") during propagation through the sandstone, indicating that the induced fractures are younger. Butts from vertical, 4-inch diameter core; uphole is towards the top of both photos.

References

Lorenz, J.C., Teufel, L.W., and Warpinski, N.R., 1991, Regional fractures I: a mechanism for the formation of regional fractures at depth in flat-lying reservoirs. AAPG Bulletin, 75, 1714–1737.

Rhett, D., 2001, Pore pressure controls on the origin of regional fractures: experimental verification of a model. AAPG Search and Discovery Article #90906, 2001 AAPG Annual Convention, Denver, Colorado.

Robinson, L.H. Jr., 1959, The effect of pore and confining pressure on the failure process in sedimentary rock. Colorado School of Mines Quarterly, 54, 177–199.

2M

Cored Hydraulic Fractures

There are numerous anecdotal reports of cores that have cut across and retrieved samples of hydraulic stimulation fractures, but there are only a few published descriptions (Fast et al., 1994; Hopkins et al., 1998; Peterson et al., 2001; Potluri et al., 2005; Warpinski et al., 1993). The characteristics of hydraulic fractures in core that are common to the several published descriptions include the following.

- Hydraulic fractures are multi-stranded with close spacing (centimeters to decimeters) (Figures 2M.1, 2M.2).
- Gel residues have been reported in the fracture apertures, but remnant proppant is rarer, the absence being attributed to washing the proppant out of the fractures during coring and processing.
- The rock may be completely separated, or it may be only cracked and remain partially intact across fracture planes.
- The hydraulic fractures show small (mm to cm) offsets at bedding planes (see Figure 2M.2).
- Some hydraulic fractures show large (meters to tens of meters) and unexpected lateral sidesteps.
- Hydraulic fractures cut new host rock more often than they exploit natural fracture apertures or break open mineralized natural fractures.
- Hydraulic fractures may strike parallel (Figures 2M.3, 2M.4) or oblique to natural fractures, depending on the orientation of the in situ stresses relative to the natural fracture strike.

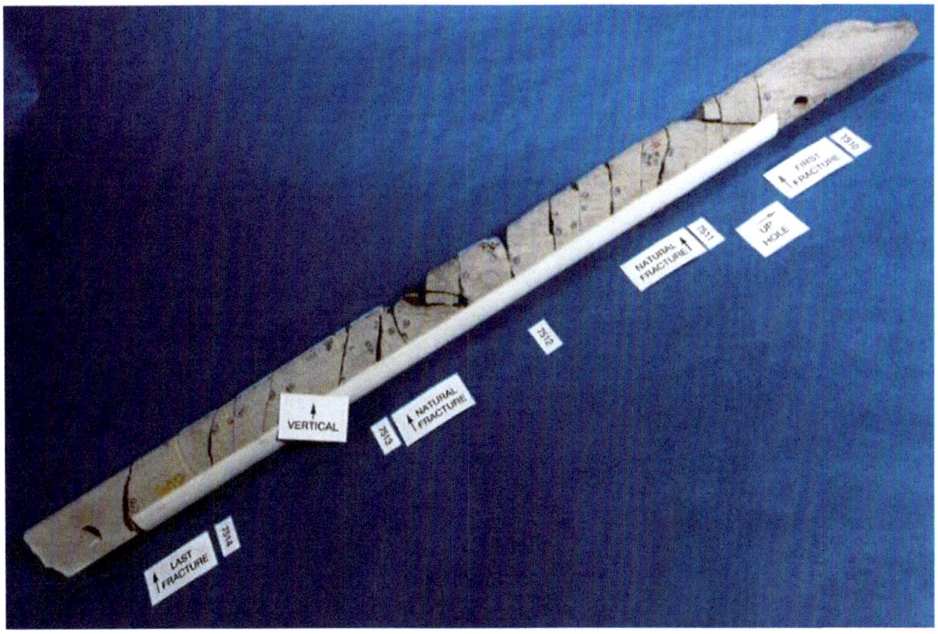

Figure 2M.1 This interval of core, cut from a non-marine sandstone in a deviated hole, intersected 29 strands of a hydraulic fracture as clarified in the following sketch. The horizontal planes are bedding planes against which some of the hydraulic fracture strands terminated and others were offset. This core is located about 50 ft from the vertical injection wellbore. Inclined, 4-inch diameter core, shown in its in situ position, with upsection towards the top of the photo and uphole towards the upper right corner of the photo. From Warpinski et al. (1993).

Atlas of Natural and Induced Fractures in Core, First Edition. John C. Lorenz and Scott P. Cooper.

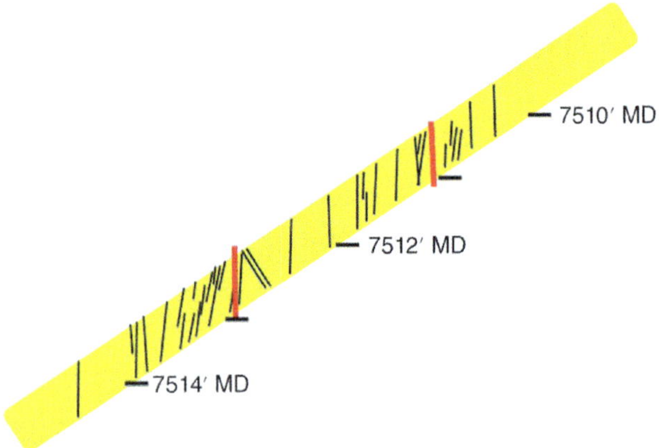

Figure 2M.2 Sketch showing the placement and approximate geometries of the hydraulic-fracture strands (black lines) shown in the previous photo. The core also cut two calcite-mineralized, nearly occluded natural fractures (red lines), which did not create mechanical weaknesses or provide apertures to divert the hydraulic fracture.

Figure 2M.3 This core, cut from a clean, moderately cemented sandstone in a deviated wellbore, shows dark, hydraulic-fracture strands ("H") that are parallel to but that did not make use of the adjacent white natural fracture planes ("N"). The natural fractures are in fact deformation bands, with no aperture and probably stronger than the adjacent host rock. The object of the core was to find out where injected, slurrified drill core cuttings go when injected into the subsurface, and it was taken specifically to characterize the hydraulic fractures (see Peterson et al., 2001). Inclined, 4-inch diameter core shown in its in situ position; stratigraphic up is towards the top of the photo, uphole towards the upper right.

Figure 2M.4 Cut from the same well as the core shown in the previous photo, this photo shows the multiple hydraulic fracture strands cutting through a marine shale. The hydraulic fractures are vertical and subparallel to each other. Inclined, 4-inch diameter core shown in its in situ position, with the obscure, weakly developed bedding cutting left to right across the photo; stratigraphic up is towards the top of the photo, uphole is towards the upper right.

References

Fast, R.E., Murer, A.S., and Timmer, R.S., 1994, Description and analysis of cored hydraulic fractures – Lost Hills Field. Dern County, California: SPE Production and Facilities, pp. 107–113.

Hopkins, C.W., Holditch, S.A., Rosen, R.L., and Hill, D.G., 1998, Characterization of an induced hydraulic fracture completion in a naturally fractured Antrim Shale Reservoir. SPE 51068, SPE Eastern Regional Meeting, Pittsburgh, PA, pp. 177–185.

Peterson, R.E., Warpinski, N.R., Lorenz, J.C., Garber, M., Wolhart, W.L. and Steiger, R.P., 2001, Assessment of the mounds drill cuttings injection disposal domain. SPE-71378-MS, presented at the SPE Annual Technical Conference and Exhibition, 30 September–3 October, New Orleans.

Potluri, N., Zhu, D., and Hill, A.D., 2005, Effect of natural fractures on hydraulic fracture propagation. SPE94568, SPE European Formation Damage Conference, Scheveningen, The Netherlands, 25–27 May.

Warpinski, N.R., Lorenz, J.C., Brangan, P.T., Myal, F.R., and Ball, B.L., 1993, Examination of a cored hydraulic fracture in a deep gas well. SPE Production and Facilities, pp. 150–158.

Part 3

Artifacts

3A

Introduction

Core artifacts are structures in the rock produced by the coring and handling processes. They are typically non-planar and do not fit into the common definition of "fracture." This book is focused on fractures, but the artifacts in core must be understood if the fractures are to be properly logged and interpreted, and some artifacts offer information useful to a fracture analysis. For example, the artifacts that can be present at the tops and bottoms of a core help in deciding whether a core that comes up short is missing an upper or lower section and thus where it fits into the stratigraphy. Artifacts such as spinoffs and core-catcher drag marks at drill pipe connections provide depth correlation points for oriented cores, which need to be depth shifted with precision relative to the core orientation report provided by the service company if the report is to be useful and valid. Moreover, an assessment of the reliability of the orientation-groove artifacts on a core surface is imperative in assessing the reliability of an orientation survey.

Other structures such as twice-turned core can provide information on the drilling and handling processes, but even features that turn out to be merely curiosities need to be recognized as such so that they can be dismissed. Determinations of the consistency of slab-plane orientations help in assessing the robustness of fracture measurements including fracture heights and strikes.

This section of the book describes a variety of the more common artifacts found in cores. Also included are more esoteric features such as the divots left on a core surface by the pinion screws used to hold the core in place within the barrel during processing.

The artifacts described in this section of the atlas include the following.

- Tops and bases of core
  - Bit impressions
  - Stumps
- Core-surface artifacts
  - Core-catcher scars
  - Scribe-line grooves
  - Irregular core diameters
  - Pinion holes
- Coring artifacts
  - Spinoffs
  - Twice-turned core
- Sawing scars
  - Chop saw cuts
  - Core-barrel-splitting scars
  - Slab saw damage
- Other artifacts
  - Rotational plucking
  - Knife scratches
  - Erosion by drilling mud
  - Polished surfaces
  - Inconsistent slabbing
  - Illusions
  - Metamorphic core rinds

Atlas of Natural and Induced Fractures in Core, First Edition. John C. Lorenz and Scott P. Cooper.

3B

Core Tops and Core Bases

When the core run is complete and the core is to be brought to the surface, the core and coring assembly need to be disengaged from the formation. As the draw works pull up on the drill string, a core catcher in the coring shoe at the bottom of the assembly tightens around the circumference of the core, fixing it in the barrel so that tension from the draw works can break the rock, ripping the core from the formation. This usually works because rock is weak in tension, especially if that tension is normal to the bedding planes in the formation.

Occasionally a bedding plane below the core catcher fails before the rock inside the catcher does, resulting in the recovery of a piece of the formation shaped like a tree stump at the bottom of the core (Figures 3B.1, 3B.2). Essentially, the bottom of the hole has been brought up with the core when the driller tried to turn the wellbore inside out.

In addition to making a nice door stop, a core stump can be important if the recovered core is significantly shorter than the cored interval, as it shows that the lost core had to have been from the top of the cored interval, being ground up before the core barrel seated rather than having been lost out the bottom of the core barrel due to failure of the core catcher, or ground up during the last part of the core run.

In the same way that core stumps mark the base of a core, drill-bit impressions (Figures 3B.3, 3B.4) indicate that the top of a core has been recovered, and they are useful in the same manner. The top of a core displaying a rotary drill-bit impression had been the bottom of the drill-hole at the time that the drill-bit was pulled up and replaced by a core bit. The core bit then captured a 4-inch diameter sample of what had been the bottom of the hole.

Bit impressions are rare within a sequence of back-to-back cores since the bottom of the hole left by a core run consists of a short core stump, although sometimes a driller will trip into a hole with a rotary drill-bit and drill ahead a few feet to clean up a hole between core runs. The impressions left by some drill-bits resemble spin-offs, but unlike spinoffs they occur at the top of a core without an opposing, mating face, and they are not marked by concentric slickenlines.

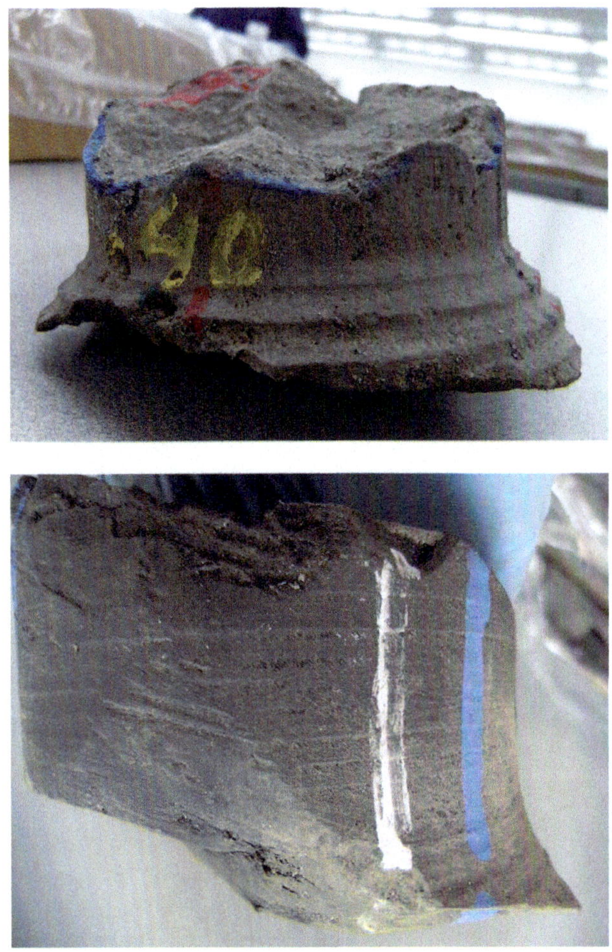

Figure 3B.1 Two stumps from the bases of cores cut from a dolomite (top) and a marine shale (bottom). The flare at the bottom of the stump was the bottom of the hole at the time coring was discontinued; the irregular bedding planes under the stumps were weaker than the overlying core and broke first when the core was pulled out of the hole. The core in the upper photo was marked with the industry standard red-black line pair to indicate uphole orientation, but the core in the lower photo uses an unconventional blue-white line pair. Vertical, 4-inch diameter cores; uphole is towards the top of both photos.

Figure 3B.2 A much smaller stump remnant (arrow) at the base of a limestone core. Even though it is a small lip, the core diameter at this location is greater than 4 inches and could not have fit into the core barrel. Core stumps can form at the bottom of any core run, and so can be found in the middle of a core box if several cores were taken back-to-back. However, they are likely to wind up on someone's desk. Vertical 4-inch core; uphole is towards the top of the photo.

Figure 3B.3 Two examples of the impressions left by toothed, 1948 Hughes tricone drill-bits at the bottoms of two wellbores when drilling stopped before the start of coring. Drill-bit impressions are artificial trace fossils of a sort, and were brought to the surface when the cores were recovered. Vertical, 4-inch diameter core; uphole is towards the viewer.

Figure 3B.4 Different types of drill-bits leave different impressions. The upper left and upper right patterns, in cores cut from a marine shale and a coarse-grained limestone respectively, were produced by PCD (PolyCrystalline Diamond) bits that scrape the formation to create smoother circular patterns at the bottom of the hole, in contrast to the hammer and spall action of the teeth of a tricone bit that creates the rough patterns shown above. The lower photo shows a low-relief but still pock-marked surface suggesting a tricone bit in a softer shale. Vertical, 4-inch diameter cores; uphole is obliquely up and towards the viewer in all three photos.

3C1

Core-Catcher Drag

A core catcher consists of steel wedges mounted inside the core shoe, just above the core bit. Different types and configurations of core-catcher wedges exist, but all are retracted into recesses inside the cylindrical core shoe as long as there is weight on the bit and everything functions correctly. Some wedges consist of a "basket" of rough-edged fingers on a split, tapered ring that fits into a corresponding taper in the shoe, others consist of separate rough-edged wedges ("dogs") that fit into tapered slots. When the driller pulls up the drill string, the wedges slide out of their housings and their rough surfaces grip the sides of a core with enough force, usually, to hold the core while the draw works are used to pull on the core and break it from the formation at the bottom of the hole.

The scars left by a core catcher where it grabbed a core will be visible on the core surface at the bottom of the core (Figure 3C1.1, left), unless, as is often the case, that last foot of core had to be shattered with a sledge hammer, the tool of choice, in order to release it from the core-catcher wedges in the core shoe. Core-catcher scars are also formed each time the driller lifts the drill string in order to screw on a new 30 ft joint of pipe as the hole deepens (Figure 3C1.1, right), although this is no longer necessary with top-drive drill rigs or downhole mud-motor operations.

The core-catcher wedges do not always retract fully into their recesses during coring, whereupon the rough wedge surfaces get dragged along the outside of the core as it is enveloped by the core barrel during coring (Figures 3C1.2, 3C1.3). The rotational position of basket-type core catchers is not fixed within the core shoe, and they may leave rotating or irregular scars on the core (see Figure 3C1.3) even if associated core-orientation scribe grooves are straight. Indistinct core catcher drag scars can be mistaken for the scribe grooves of an oriented core. Some scars left by core catcher drag can be quite deep, leaving distinct, concave-downward chatter marks (Figure 3C1.4).

Atlas of Natural and Induced Fractures in Core, First Edition. John C. Lorenz and Scott P. Cooper.

Figure 3C1.1 Left: the scars left by individual core-catcher wedges or "dogs" (red arrows) where they clamped the base of a core, and the similar scars made by a core orientation scribe (black arrow). The catcher wedges were not fully retracted and were dragging along the core surface above the engagement point. The catcher and scribe assembly rotated rapidly around the core, creating spiral scars, and the core orientation survey would be invalid for this section of the core. Right: the scars left by a different, basket type of core catcher consisting of multiple fingers. This scar pattern was made at a connection rather than at the bottom of the core, as the core below the scars is a continuation of the same core run. Core orientation scribe lines usually jump at connections, providing a depth correlation point to the orientation survey. Both photos are of vertical, 4-inch diameter cores, with uphole towards the top of the photos.

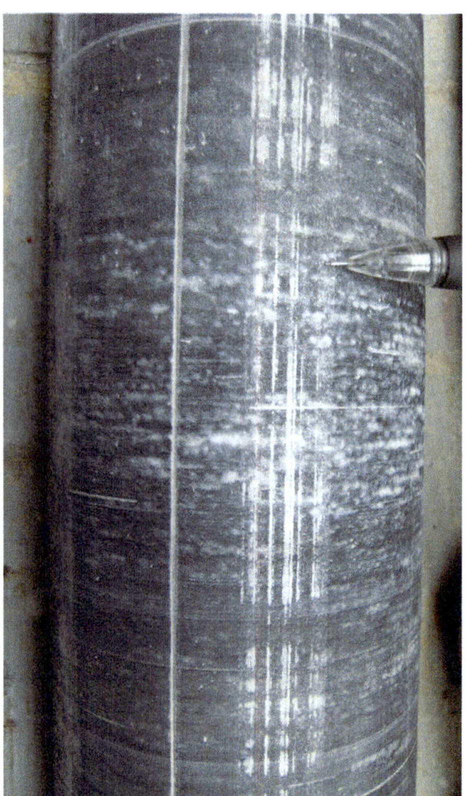

Figure 3C1.2 One of the individual core-catcher dogs did not fully retract while coring this limestone, and dragged along the core surface, leaving a broad scar (immediately to the left of the pencil). This scar contrasts with the deeper, more discrete core orientation scribe groove to the left of the core-catcher scar. Incomplete catcher retraction usually does not significantly damage a core or impede coring operations, but if the catcher is stuck and fully extended the core cannot be cut. Vertical, 4-inch diameter core; uphole is towards the top of the photo.

Figure 3C1.3 Scars left by incompletely retracted multi-finger, basket-type core catchers during coring of a limestone (left) and a calcareous shale (right). The basket rotated clockwise downhole in the example on the right, being dragged by the rotation of the drill string. The basket in the example on the left was more completely isolated from the rotating drill string and left scars that did not spiral, although the vibration inherent in the mechanics of coring and drilling caused some wandering. Both cores are vertical and 4 inches in diameter; uphole is towards the top of both photos.

Figure 3C1.4 Incompletely retracted catchers can damage cores cut from less competent rock such as the muddy shales shown here. The malfunctioning wedges were dragged along the core surface as the core barrel was lowered over the core, creating crescentic chatter marks that are concave downhole. Some such scars extend for many feet along the core, ending where the catcher finally retracted or where the lithology became more competent. Vertical, 4-inch diameter cores; uphole is towards the top of both photos.

3C2

Core Orientation Scribe Grooves

An oriented core is scribed by three "knives," triangular wedges of hardened steel that are mounted with the core catcher inside the core shoe, which is screwed to the base of the core barrel inside the core bit. The scribe knives are positioned around the inside of the shoe so that as the core enters the barrel during coring, the knives scratch millimeter-deep grooves at three positions on the core surface. The grooves are typically but not always arranged at 0°, 150°, and 230° relative to each other around the core looking downhole at the top of the core.

The 0° knife, referred to as the "toolface" in an orientation report, cuts the principal scribe line, a groove which is no deeper than the others but which is recognized by being set apart from the pair of more closely spaced secondary grooves. Different service companies use different scribe positions, but an asymmetric configuration using the 0°-150°-230° positions not only orients the core but also cuts a unique up-down orientation reference onto the core. Moreover, if one of the scribe knives fails or a scribe groove is removed during slabbing, this configuration allows a determination of the position of the principal scribe with only two scribe lines.

The service companies provide a core orientation report that lists the position of the principal scribe line clockwise from north at each foot of depth (or, for a horizontal core, clockwise from up when looking downhole). An orientation report rarely indicates the configuration of the scribe knives in the shoe, which must be measured from the core.

The trick is that the depth reported in a core orientation report is the driller's depth, and it *must* be depth-shifted to exactly match core depth when calculating fracture strikes since the core scribe lines rotate with depth. Ideally, the scribe shoe and inner barrel are isolated by bearings from the rotating core bit and barrel and should not rotate, but there is enough friction in the system that rotation is common. Straight scribes (Figure 3C2.1) are the ideal, but rotations of the scribe grooves by up to about 10° per foot (Figure 3C2.2) are usually acceptable. Greater amounts of rotation become increasingly difficult to work with since an error of a few feet in correlating core depth to the orientation survey depths leads to significant changes in the calculated fracture strikes.

For this reason, it is imperative to recognize artifacts such as connections, spinoffs, scribe shifts (Figure 3C2.3), and scribe jumps (Figure 3C2.4) in a core, as these discontinuities in the principal scribe line should match jumps in the reported toolface orientation in the orientation report, providing depth correlation points. Moreover, the per-foot degrees of rotation of the principal scribe line on a core surface should match the reported Toolface rotation; there is a problem that must be recognized and resolved if the two are not compatible.

A straight, master orientation line (MOL) drawn on continuous lengths of core (usually drawn in green; Figure 3C2.5) can provide a reference for measuring scribe line deviation by foot, and sometimes scribe rotations relative to a MOL provide the only means of depth-shifting a core to an orientation survey. Variations in the rotation of scribe grooves with depth can and should be shifted to match variations in toolface rotation reported by the orientation company. If a core cuts slowly or if there is excessive vibration in the system, the scribe knives can cut excessively wide grooves in the core (Figure 3C2.6), leading to 15–20° uncertainties in reconstructed fracture strikes. Such wide grooves can resemble core-catcher drag.

Some companies prefer not to orient cores since the stresses imparted by scribe knives can break up a core, but oriented cores can successfully be cut from many fractured formations. Considerations include degree of fracture mineralization (strongly mineralized fractures are not as easily broken open by the scribe knives), length of the cores to be cut (fractures at the base of tall cores can be more easily broken open by knives due to the weight of the overlying core), and the competency of the rock being cored.

Atlas of Natural and Induced Fractures in Core, First Edition. John C. Lorenz and Scott P. Cooper.

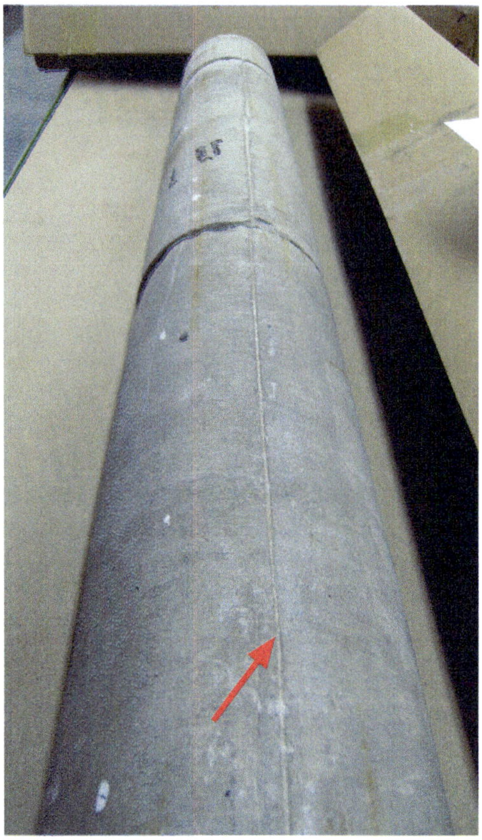

Figure 3C2.1 The ideal scribe line (arrow) is nearly straight, rotating very little or not at all down the axis of a core. Vertical, 4-inch diameter core; uphole is away from the viewer.

Figure 3C2.2 Most scribe lines spiral clockwise around a core due to friction between the rotating outer and supposedly stationary inner core barrels. Occasionally, a scribe rotates counterclockwise, probably due to vibrational harmonics in the drill string. The scribe lines in this photo (arrows) rotate at a uniform rate for a total of about 80° of rotation around the core axis along the 8-ft interval shown, or about 10° per foot. If there is a plus or minus 1-ft depth uncertainty in correlating a fracture in the oriented core to the orientation survey, the calculated fracture strike is only accurate to plus or minus 10°. Scribe line rotation cannot be properly measured unless the core is reassembled and locked together. Few service companies currently offer this assessment. It is a critical, though often omitted, step when measuring fracture strikes from an oriented core. Note the PCD (PolyCrystalline Diamond) bit impression at the top of the core. Vertical, 4-inch diameter core; uphole is towards the viewer.

Figure 3C2.3 Scribe line rotation is not always uniform; it can shift abruptly, sometimes due to changes in lithology, sometimes due to changes in the drilling parameters such as weight on bit or drill string rotation rate. In this photo, the scribe knives were abruptly dragged sideways around the core surface by 40–50°, providing a measurable scribe line shift that should be reflected in the toolface orientation report. This marks a depth correlation point between the report and the core. Vertical, 4-inch diameter core; uphole is towards the top of the photo.

Figure 3C2.4 In both horizontal and vertical cores, orientation scribe grooves may jump to a new position around the core at a natural or induced break even though the core pieces lock end to end. These scribe skips may be related to minor bounces in the bit during coring, and again provide depth correlation points between core depths and the orientation survey. Both cores shown here are horizontal and 4 inches in diameter. The upper core is from a marine limestone, the lower one from a marine siltstone; uphole is to the left of both photos.

Figure 3C2.5 A green master orientation line (white arrow) is useful whether or not a core is oriented. Drawn straight down the axis of a core for each continuous section and over lockable breaks in the core, such a line can be used to measure the rotation of a scribe line (red arrow) with depth (note how the scribe line rotates to converge with the green orientation line). A master orientation line also provides an orientation reference for structures relative to each other above and below subsequently removed samples of core. However, a master orientation line is useless unless it is accompanied by a record of the discontinuities in the line, at rubble zones, spinoffs, etc. The photo on the right shows where a service company was unclear on the concept, marking a green master orientation line through bagged rubble. Horizontal core; stratigraphic up is at the high side of the core as it sits on the table, and uphole is away from the viewer.

Figure 3C2.6 Vibrations inherent in the drilling and coring machinery can be transmitted to the core orientation mechanism, and can result in poorly defined scribe line grooves in less competent strata. In addition to an excessive degree of scribe rotation along this core piece, the scarring of this core surface by a vibrating set of scribe knifes created a distressed "groove" that is 10–15° wide, meaning there will be a similar error bar on any fractures measured relative to the groove at this depth. Vertical, 4-inch diameter core; uphole is towards the top of the photo.

3C3

Irregular Core Diameters

Ideal cores have a constant diameter along their lengths, but a number of processes can result in irregular and/or smaller core diameters. Excess drill string weight on a core bit can bend the core barrel just enough to tilt it, the azimuth of tilting rotating slowly around the bottom of the hole as the drill string rotates more rapidly, milling the sides of the core (Figures 3C3.1, 3C3.2). A core surface can be milled to a smaller diameter by dragging core orientation scribe knives sideways around the core circumference if the formation is soft, or if the bearings that isolate the inner from the outer core barrel fail. Short sections of small-diameter, necked core are also common at the top of a core where the bit has not yet seated itself into a circular kerf at the bottom of the hole (Figures 3C3.3, 3C3.4).

Figure 3C3.1 Excess weight on the core bit, bending the core barrel and tilting the bit slightly sideways in the hole, was probably responsible for the spiraling channels worn into the sides of this core, cut from an oil-soaked eolian sandstone. The azimuth of tilt rotated around the wellbore as the hole deepened, resulting in the downward and clockwise spirals. Vertical 5-inch (and heavy) diameter core; uphole is towards the top of the photo.

Figure 3C3.2 Two views of necking in the middle of a core cut from an oil-soaked sandstone. There are no spirals or other clues to the origin of the necking, but excess weight on the bit resulting in core bit tilting is a plausible mechanism. The arrow drawn on the core in the right photo is of unknown significance, but points downhole. Vertical, 3.5-inch diameter core; uphole is towards the top of both photos.

Atlas of Natural and Induced Fractures in Core, First Edition. John C. Lorenz and Scott P. Cooper.

Figure 3C3.3 A reduced-diameter section at the top of a core, made by wobble in the rotating core bit at the bottom of the hole before it seated. Vertical, 4-inch diameter core; uphole is towards the top of the photo.

Figure 3C3.4 The scribe knives used when orienting this core were dragged sideways around the core, scraping the surface and reducing the diameter of the core by several millimeters until they finally formed a groove (arrow) capable of stabilizing the core shoe and inner core barrel. Note the drill bit impression at the top of the core, indicating that the core captured the precore bottom of the wellbore. Vertical, 4-inch diameter core; uphole is towards the top of the photo.

3C4

Pinion Holes

Some operations use bolts screwed into and through the side of a core barrel to pinion the core in place, preventing it from shifting or falling out of the barrel as the coring assembly is disassembled and the inner barrel is processed. The bolts leave holes in the core (Figures 3C4.1, 3C4.2) and can create local spalls and fractures around the holes, scarring that can usually be traced back to the bolthole origination point.

Figure 3C4.1 Two pinion boltholes on the sides of a core, scars from the bolts used to stabilize the core in the core barrel during processing. The diagonal scratch upper left to lower right is probably from a dragging core catcher. Horizontal, 4-inch diameter core; uphole is to the left.

Atlas of Natural and Induced Fractures in Core, First Edition. John C. Lorenz and Scott P. Cooper.

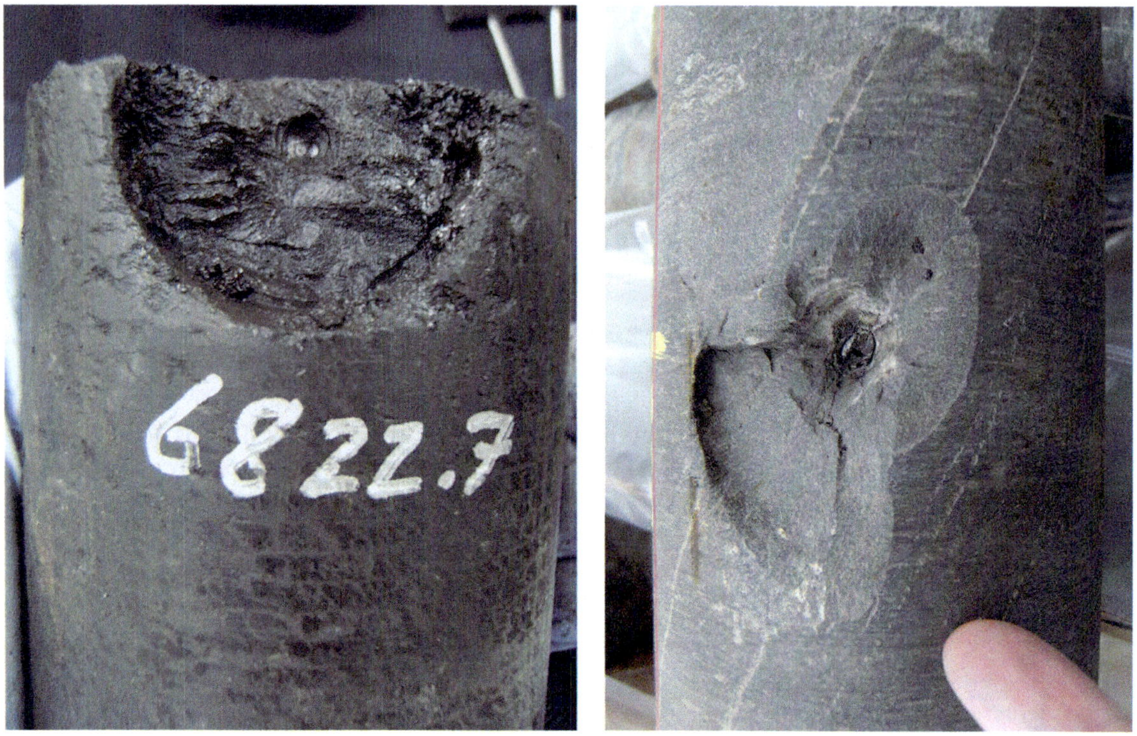

Figure 3C4.2 These cores, cut from a muddy marine shale (left) and a dolomite (right), spalled and fractured around pinion boltholes. The bolts can create irregular fractures as they are forced into the core. Vertical, 4-inch diameter core; uphole is towards the top of the photo.

3D1

Spinoffs

When the outer core barrel is imperfectly decoupled from the inner barrel during coring, the inner barrel starts to rotate, imparting a torque to the core in the barrel. If the base of that core is still attached to the formation, small amounts of torque will form the induced torque fractures described in section 2F. If a core captures core-normal weakness planes such as bedding, or if larger amounts of torque are imparted to the core by excessive inner barrel rotation, torque may not only break the core but may also cause the core above that break to rotate with the drill string and spin against the core that is still attached to the formation, creating a spinoff plane. Spinoff surfaces display distinctive concentric circular lineations that were formed by the continuous rotation of one core surface against the other. Slickensides may even form where rotation occurred over a long period of time.

Spinoffs may be planar, they may be concave upwards or downwards (Figures 3D1.1–3D1.3), or they may be a combination of all three (Figure 3D1.4).

Pseudo-spinoffs can occur in horizontal core, where system vibration during coring caused two core pieces to rock back and forth against each other without continuously spinning in the direction of drill-string rotation (Figure 3D1.5).

Spinoffs are common, and some cores are broken by numerous regularly spaced spinoffs. In oriented cores they can be important depth correlation structures if they are not so abundant that more than one might be correlative to jumps in the reported toolface orientation. Some spinoffs represent the loss of a considerable volume of rock, ground up during rotation of the two spinoff faces against each other.

Atlas of Natural and Induced Fractures in Core, First Edition. John C. Lorenz and Scott P. Cooper.

Figure 3D1.1 Two views of a spinoff in a limestone. The upper photo shows the irregularly congruent concave-convex surfaces of rock above and below the spinoff. The lower photo shows one of the spinoff surfaces, marked by subtle concentric lineations. Vertical, 4-inch diameter core; uphole is towards the top of both photos

Figure 3D1.2 Irregularly concentric lineations mark the planar spinoff surfaces in core from a marine shale (top) and in a vuggy dolomite (bottom). The concentric nature of the lineations in the upper photo is obscured by smears left when wet drilling mud was wiped off the core with a rag. Vertical, 4-inch diameter cores; uphole is away from the viewer in the upper photo, and to the right in the lower photo.

Figure 3D1.3 An irregular spinoff in a siliceous shale (left) and in a slab segment from a marine shale (right), showing concentric lineations indicating circular shear of one piece of core against another.

Figure 3D1.4 Two views of a locking spinoff (arrow, left photo). The right photo shows the opposing faces that display the characteristic concentric circular pattern of a spinoff. The surfaces are not planar but rather are three-dimensional and fully congruent to each other. The two core faces cannot move laterally relative to the core axis but rotate easily against each other. In one sense, the core "locks" across this spinoff, but it is free to rotate so it provides a good depth correlation point to the jump in the reported toolface orientation for this oriented core. The green master orientation line, intended to indicate continuous locking core intervals and provide a reference in those intervals for relative fracture strikes, has been mistakenly drawn across the spinoff without the arrowheads indicating interruption. Vertical, 4-inch diameter core, uphole is towards the top of the left photo; it is away from the viewer for the upper core piece and towards the viewer in the lower core piece in the right photo.

Figure 3D1.5 A view of the opposing faces of a circular structure in horizontal core from a muddy limestone. Remnants of arcuate fracture-propagation ribs that propagated around the circumference of the core (arrow) show that the original break across the core was a disc fracture (see section 2D), which was subsequently exploited as a pseudo-spinoff. Close inspection shows that the concentric lineations on the spinoff surface are short arcs rather than continuous circles, suggesting back-and-forth rotational oscillations due to vibration rather than continuous spinning of the two faces against each other. Horizontal, 4-inch diameter core; uphole is towards the viewer in the left core piece, away from the viewer in the right core piece.

3D2

Twice-Turned Core

Some core pieces show evidence for having been cored at two or more angles. In some cases, the core bit re-entered the hole in a slightly different position between one core run and the next (Figures 3D2.1, 3D2.2), in other examples a piece of core, perhaps dropped to the bottom of the wellbore after the core catcher slipped during core retrieval, was caught at an odd angle when the core bit reached bottom again, ready for the next core run (Figure 3D2.3).

Figure 3D2.1 A short stump of core may be left in the bottom of a wellbore at the end of a core run. The stump may not be centered under the core bit when it re-enters the hole for the next run. The structure shown in this photo, recovered at the top of an intermediate core run in a series of back-to-back cores, is the remnant of an off-center core-base stump. Vertical 4-inch diameter core; uphole is towards the top of the photo.

Atlas of Natural and Induced Fractures in Core, First Edition. John C. Lorenz and Scott P. Cooper.

Figure 3D2.2 This piece of core has a shape that would seem to be impossible given the entirely circular motion of the cutting tools. In fact, it is similar to the half core stump shown above, formed by two, non-concentric passes of the core bit over the same piece of rock. Pieces like this occur at the top of a core run or where the core bit was lifted and set back down. Vertical core, initially 4-inch diameter; uphole is towards the top of the left photo and towards the viewer in the right photo.

Figure 3D2.3 Three views of a piece of core that fell to the bottom of a wellbore after being lost out the bottom of a core barrel and that was spun around in several positions under the core bit, forming a dished surface at the top of the underlying core as the bit seated for the next core run. The core fragment was finally caught by the core barrel, being recovered at the top of the core run. Vertical, 4-inch diameter core; uphole is towards the top of the left and lower right photos, uphole is unknown for the broken multi-turned core piece.

3E

Saw Scars

Masonry saws are probably the second most common tool, after hammers, used in core processing. Core-related uses for saws include cutting cores to length for transport and boxing, removing cores from inner aluminum barrels, cutting cores for sampling, and slabbing.

Proper use of saws requires finesse. Slab saw blades must be periodically dressed to keep the diamonds embedded in the cutting edge exposed, and saw blades should not be forced through a core when slabbing. Proper slabbing technique can produce a smooth slab face whereas poor slabbing technique and a dull blade leave scars on the slabs that must be removed by sanding. Sanding adds wear and tear on the core, and produces rock flour that gets jammed into all available cracks in the slab and often resembles mineralization.

When cutting a core into lengths for boxing or transport, the core may be scored, i.e., sawn part-way through, and then broken the rest of the way. This process has the advantage of leaving the core ends lockable and the segments at their true lengths. However, the incomplete saw cuts (Figures 3E.1, 3E.2) are often intentionally positioned in the core butts during slabbing, and, if only the slabs are examined, the regular induced fractures related to scoring can resemble natural fractures, especially in horizontal cores where these core-normal saw-related fractures are vertical. More often, the cores are scribed around their circumference before being broken (Figure 3E.3), or even sawn all the way through, especially when they are being cut into transportable lengths while still inside the core-barrel liner on site at the drill pad.

Some service companies have started to cut core into short, 6-inch lengths that are easier to handle and that more readily fit into the slabbing equipment. Although the stratigraphy and most sedimentary structures can still be adequately assessed in such core fragments (except in a horizontal core), this is not optimum for studying fractures as it increases the potential for non-parallel slab planes and inverted core pieces.

When core-barrel liners first came into general use in coring operations, there was no consensus on the best way to remove core from a liner, and one can find longitudinal saw-cut nicks on cores from this era, where the core was cut out of the liner by sawing the liner in half lengthwise (Figure 3E.4). This was both labor intensive and dangerous and has largely been discontinued.

Slab saws are another, not always obvious source of unusual artifacts. Dull blades pushed through a core with too much force can produce scars ranging from etched arcs on a slab face (Figure 3E.5) to arc-shaped pieces of core, still smelling of the overheated saw blade and burnt rock, spalled off when the technician pushed and twisted the core on the blade (Figures 3E.6–3E.8).

Even with the best technique, when a fracture is oblique to the plane of slabbing, slab saws can damage the thin wedge of rock caught between the fracture and the slab plane (Figures 3E.9, 3E.10). The damage zone resembles mineralization and makes fractures to be appear much wider than they are.

Photographs in this section are presented with the core axis horizontal since that is the position in which the cores are processed and sawn, whether they were cut from vertical, inclined, or horizontal wells. It is easier to understand the marks left from saws when the core is viewed in the position in which it was cut.

Atlas of Natural and Induced Fractures in Core, First Edition. John C. Lorenz and Scott P. Cooper.

Figure 3E.1 Some cores are sawn half-way through with the expectation that it will be easy to break them the rest of the way across the core, and this usually works but it is not infallible. The upper photo shows core that has been sawn part-way and broken the rest of the way, the saw cut being the obvious origin for the break in the core. The lower photo shows core that did not behave as expected, with an induced core break *between* the closely spaced saw cuts. The art of breaking a core at a desired location with a hammer has largely been lost. Top: butt from a horizontal core; stratigraphic up is towards the top of the photo, uphole is to the right. Bottom: butt from a vertical core; uphole is to the left (note the scars from a rotating core catcher).

Figure 3E.2 Two views of a horizontal core that was cut half-way through with a saw at regular, closely spaced intervals, and then broken the rest of the way at each saw cut. Top photo shows a saw cut in the core butt, viewed end-on; the saw cut was not deep enough to intersect what later became the slab plane. Bottom photo shows slab pieces from the same core, containing only the breaks across the core, which would be of uncertain origin without having noted the associated saw cuts in the core butts. Slabs and butts of horizontal, 4-inch diameter core; stratigraphic up is to the right in the upper photo and towards the top of the lower photo; uphole is towards the viewer in the upper photo and to the right in the lower photo.

Figure 3E.3 These two cores, cut from a limestone (*top*) and a marine siltstone (*bottom*), were scored around their circumference with a saw, cutting through the aluminum core barrel liner and nicking the core. Neither core broke at the incomplete saw cut as intended. The upper photo is vertical, 4-inch diameter core; uphole is to the left. The lower photo shows the contact between a siltstone and a black shale in a horizontal 4-inch core; stratigraphic up is away from the viewer and uphole is to the left.

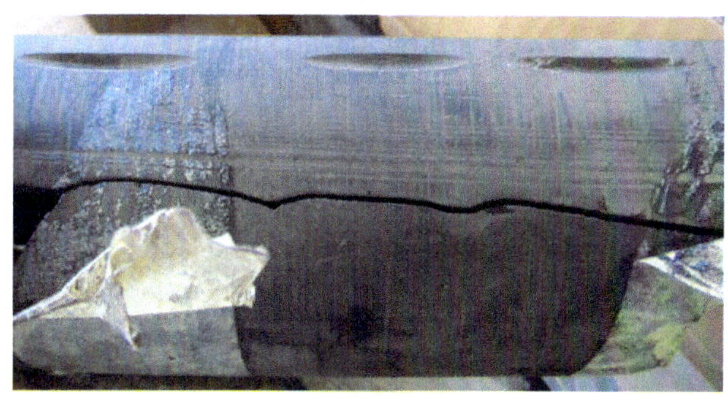

Figure 3E.4 The three dashed saw cuts along the upper surface of this core were made as the core barrel liner was split longitudinally in order to remove the core. The use of aluminum core barrel liners increases the chance of recovering a core intact and decreases the potential for core jamming, but it has its own problems. Numerous techniques have been used for removing core from liners, including plungers, car jacks, and pumping it out with the rig mud pump (the potato cannon technique). Horizontal, 4-inch diameter core; uphole is to the left.

Figure 3E.5 Poorly maintained and improperly used slab saws will scar a slab face with an arcuate record of the saw passage. This oil-soaked sandstone core became misaligned with the slab saw blade part-way through the slab cut, resulting in these slab saw scars. Slab from a vertical, 4-inch diameter core; uphole is to the left.

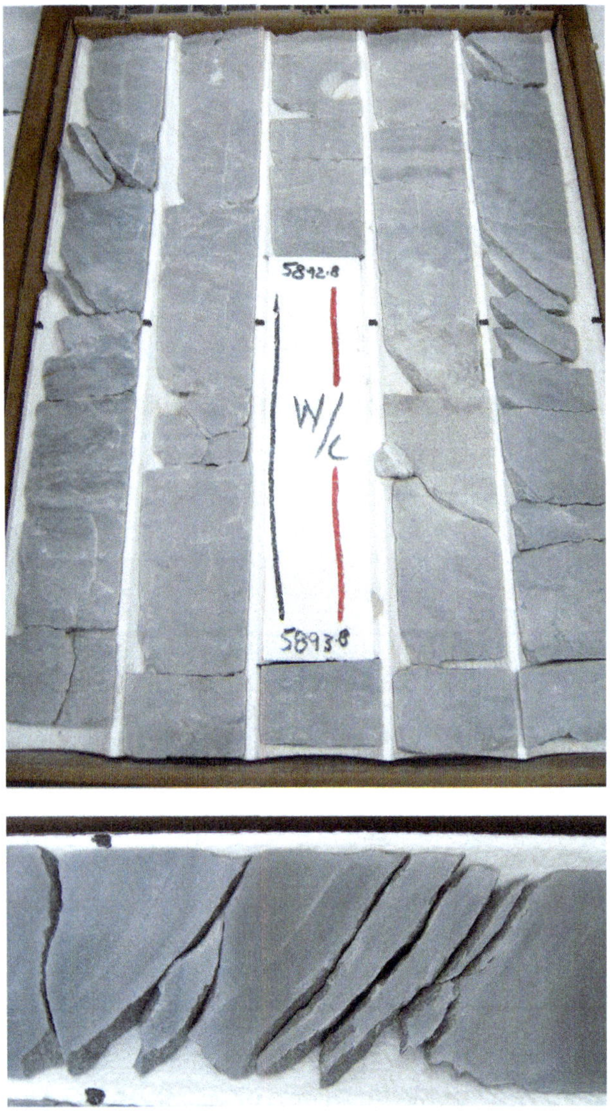

Figure 3E.6 Two examples of arcuate induced fractures caused by slabbing. Technicians sometimes force a slab saw through the core, turning arcuate slab saw scars into arcuate slab saw fractures. The relationship between slabbing and the arcuate fractures may not be obvious since the saw scars on the slabs have usually been removed by sanding, but the diagnostic scars can be found on the opposing, unsanded butt slab face. Both of these photos are of the slabs of vertical 4-inch diameter cores cut from marine limestones; uphole is towards the top of the upper photo, and to the right of the lower photo.

Figure 3E.7 Two examples that compare slabs to butts over the same core intervals in a recrystallized limestone (*top*) and a muddy calcareous grainstone (*bottom*). The absence of arcuate slab saw fractures in equivalent core butts, as well as the presence of unsanded arcuate slab saw scars on the butt slab faces, confirms the saw-related origin of the fractures. Both cores are vertical, 4-inch diameter core, and uphole is to the right.

Figure 3E.8 Some slab saws have been forced through the core with so much pressure that the core still reeks of overheated machinery, and the arcuate slab saw fractures have twisted planes (*arrows*) created when the slab saw blade was essentially used to pry the slab fragments from the core, twisting the core. Vertical 4-inch diameter core cut from a muddy limestone; uphole direction unknown.

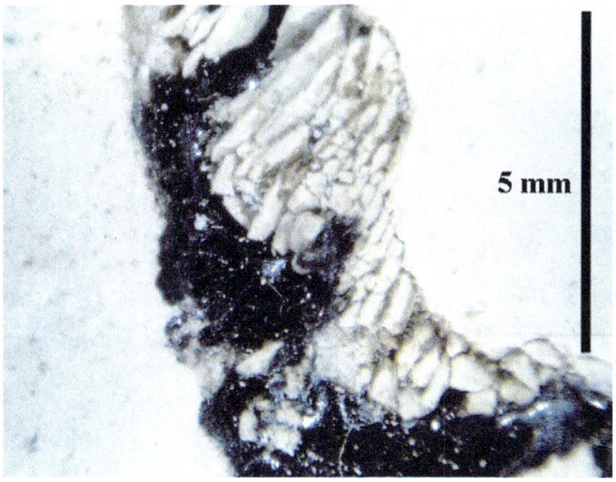

Figure 3E.10 This close-up photo shows the shattered wedge of limestone created when the slab saw cut at a shallow angle oblique to a natural fracture plane filled with a tarry bitumen. The bitumen helped hold the shattered rock in place. The slab plane is parallel to the plane of the photo, the natural fracture plane angles into the photo to the right at about a 10° angle to the slab surface. Vertical 4-inch diameter core; uphole is towards the top of the photograph.

Figure 3E.9 The narrow edge of rock between a fracture plane and a slab surface can be damaged by a slab saw or by the sanding process, leaving a zone of cracked rock on one side of the fracture plane that often resembles mineralizaiton. These two views of a damage zone along an induced fracture are from a marine limestone. Slabs from a vertical, 4-inch diameter core; uphole is towards the top of both photos.

3F1

Core Plucking

Core barrels rotate clockwise (looking downhole) during coring, so one of the two edges of any natural or induced fracture that is unsupported by tight mineralization will be subjected to tension as the bit rotates across the cross-section of the fracture plane, and the opposing face is subjected to compression.

Rock is an order of magnitude weaker in tension than it is in compression, therefore the fracture edge under tension commonly becomes plucked and irregular, whereas the opposing edge under compression remains relatively straight and clean (Figures 3F1.1, 3F1.2). This allows a geologist to determine top from bottom of a fractured core piece without recourse to the up-down core orientation markings.

Figure 3F1.1 Two sides of a vertical induced centerline fracture plane viewed edge-on in a limestone core. The left edge is straight and clean since the stress created by the rotating bit (the direction of rotation indicated by the arrow) was compressive and rock is strong in compression. The opposing fracture edge has been badly chipped and plucked since the stresses imparted by the core bit as it passed by this unsupported surface were tensile. The fracture edge on the other side of the core shows similar asymmetric plucking. Vertical, 4-inch diameter core; uphole is towards the top of the photo.

Figure 3F1.2 Natural fractures can also be plucked by a rotating bit, as shown here by the differential plucking on two sides of a dissolution-enhanced, calcite-lined fracture plane in a calcareous shale. The curved arrow drawn on the core shows the direction of bit rotation across the core surface, while the top half-arrow indicates a horizontal shear plane along a thin bentonite. Vertical 4-inch diameter core; uphole is towards the top of the photo.

3F2

Scratches

There are many opportunities for a core to be scratched during coring and processing, and although it sounds improbable, scratches can and have been mistaken for narrow, mineralized fractures.

A common reason for a linear, fracture-mimicking scratch is a knife point (usually a box cutter) intentionally drawn along a core surface in order to quickly remove the plastic film that was used to retard dehydration of a core immediately after it was cut. The film is only effective for a few weeks, so when the core needs to be examined later, technicians draw a knife along the core, cutting through the film and leaving a linear scratch on the outer core surface (Figure 3F2.1). The film is commonly cut along at least two surfaces of the core for ease of removal, meaning that many such pseudo-fractures can even be traced in three dimensions, adding to the impression that they are natural fractures (Figure 3F2.2). Scratches can leave a white trace that looks like mineralization. Examination with a hand lens or binocular microscope usually resolves the issue.

Figure 3F2.1 A subtle scratch across the end of a core cut from limestone has the appearance of a narrow, mineralized natural fracture, but was created by a knife point drawn across the core. Fragment of a vertical, 4-inch diameter core; uphole is away from the viewer.

Figure 3F2.2 The faint scratch between the arrows can be traced continuously for a foot along the core surface as well as up to and across the end of the core. The scratch was created by removing the plastic film wrap from the core with a knife. Examination with a hand lens shows that what appears to be mineralization is rock damaged by the point of the knife, and that the damage does not extend into the small valleys between the ridges created by bit passage that ring the core surface. Vertical 4-inch diameter core cut from a calcareous shale; uphole is towards the top of the photo.

Atlas of Natural and Induced Fractures in Core, First Edition. John C. Lorenz and Scott P. Cooper.

3F3

Drill-Mud Erosion

Some fracture surfaces appear to be smoothed, grooved, and eroded by high-pressure fluid flow within the fracture aperture (Figure 3F3.1). The grooves are typically parallel to the core axis, and are probably related to high-pressure flow of drilling mud within the fracture aperture and related to lost circulation into the formation. This erosion of the fracture faces is not common, but has been observed along both natural and induced fracture planes.

Figure 3F3.1 The surfaces of some centerline fractures appear to have been eroded and grooved by the rapid flow of high-pressure drilling mud along the fracture plane, as in this core cut from a deep-marine siltstone. See also Figure 2H.5. Vertical, 4-inch diameter core; uphole is towards the top of the photo.

3F4

Core-Parting Enigmas

In rare examples, cores exhibit seemingly impossible overlapping core-parting configurations. In these enigmatic structures, the core pieces on either side of an irregular parting unequivocally lock together. The core sections above and below the locking parting are both full-diameter cores, but the axes of the two core pieces are offset by up to half of the core diameter (Figure 3F4.1). It is difficult to envision how the core bit cut the upper core piece without marking the lower core piece, and a similarly ambiguous process kept the upper core piece out of the way while the lower piece was being cut.

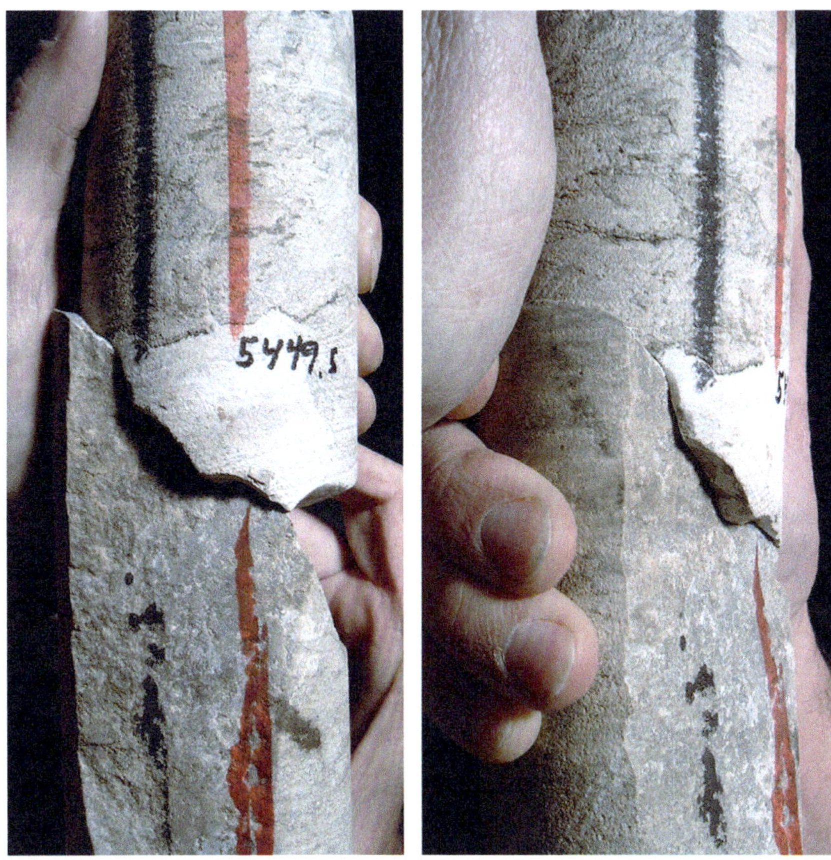

Figure 3F4.1 Two views of an enigmatic, locking connection in core from a sabkha dolomite. The core is full diameter above and below the connection. The two core pieces overlap and they are not coaxial. The lower core section shows no record of the cutting of the upper core piece even though the core pieces overlap, and the upper core section was not affected when the lower piece was cored. Vertical 2.5-inch diameter core; uphole is towards the top of both photos.

Atlas of Natural and Induced Fractures in Core, First Edition. John C. Lorenz and Scott P. Cooper.

3F5

Polished Fracture Surfaces in Horizontal Cores

The faces of both natural and induced fractures in horizontal cores sometimes display smooth, polished surfaces, but without the circular lineations that characterize spinoffs. These polished surfaces form in horizontal cores when the core parts along a fracture plane during coring, and the two faces of the parting vibrate against each other without rotation while the rest of the core is being cut (Figures 3F5.1, 3F5.2).

The fit of the core across these polished surfaces is commonly imperfect, suggesting that some rock was lost during the polishing process. Remnants of mineralization are sometimes found in hollows on the fracture faces, but polishing the core ends can remove fracture-face fractography and mineralization (Figure 3F5.3).

Figure 3F5.1 Two views of a partially polished, planar natural fracture, striking oblique to the axis of a horizontal limestone core. The left photo shows the fracture as it appears on the slab face of the core, the right photo shows the fracture plane opened to reveal one of the fracture faces. The upper half of that face has been mechanically smoothed and polished, the lower half retains remnants of the original rough fracture-face texture. Butts from a horizontal, 4-inch diameter core; up is towards the top of both photos, uphole is to the right.

Atlas of Natural and Induced Fractures in Core, First Edition. John C. Lorenz and Scott P. Cooper.

Figure 3F5.2 Two views of a polished, nearly core-normal fracture face in a core cut from a marine siltstone. Hints of remnant fracture-face mineralization in the hollows of the fracture suggest that it is natural, but the evidence is not conclusive. Horizontal, 2.5-inch diameter core; in the upper photo, uphole is to the left, stratigraphic up is unknown but immaterial; in the lower photograph, uphole is towards the viewer.

Figure 3F5.3 Two views of polish on a vertical natural fracture surface in a horizontal core cut from a limestone. Vibration has polished and removed mineralization from part of the fracture surface. The oblong shape of the fracture face is the result of the oblique angle at which the fracture cuts across the core. Horizontal, 4-inch diameter core; stratigraphic up is unknown, uphole is towards the upper right in the top photo, and towards the viewer in the bottom photograph.

3F6

Tip Polish

The ends of core pieces from a horizontal core that has broken along core-oblique fracture planes are angled rather than squared off. The tips of the opposing wedge-shaped core ends point uphole and downhole, and may be truncated by a centimeter or two for what seems like no apparent reason (Figure 3F6.1). The small, core-normal truncation surfaces are marked by radial patterns similar to those found on spinoffs (Figure 3F6.2) even though there seems to be no opposing, similarly marked face. The pattern can be repeated several times along a core, and the truncated wedge tips can be observed in CT scan images taken before the core is removed from the liner (Figure 3F6.3).

This pattern forms when a horizontal core has parted along the oblique fractures during coring, producing core pieces shaped like trapezoids in plan view (Figure 3F6.4). The unconfined core pieces are free to slide up the horizontal barrel where gravity causes the longer, heavy sides of the trapezoids to rotate around the axis of the core and settle on the low side of the vibrating barrel.

When this happens, the uphole and downhole wedges meet tip to tip and vibrate against each other (Figure 3F6.5), grinding away the tips of the core and leaving short, concentric arcs on the opposing surfaces. When the core is brought out of the hole, the core pieces are repositioned by gravity to their original positions across the fractures, and the fact that the fractures had parted, allowing the tips of the core wedges to meet while they lay horizontally in the core barrel at depth, is not obvious.

Atlas of Natural and Induced Fractures in Core, First Edition. John C. Lorenz and Scott P. Cooper.

Figure 3F6.1 Three views of a pair of calcite-mineralized, vertical fractures cutting oblique to the axis of a horizontal core, showing the missing uphole and downhole tips of the rock wedges adjacent to the fractures. Horizontal, 4-inch diameter core; stratigraphic up (the high side of the core) is towards the viewer, uphole is to the left.

Figure 3F6.2 View of the same core piece end-on across the fracture plane, showing the remnants of the mineralized fracture face (lower two-thirds of the photo) and the surface of the missing tip of the core wedge, marked by concentric arcs (arrow). Horizontal, 4-inch diameter core; uphole is away from the viewer.

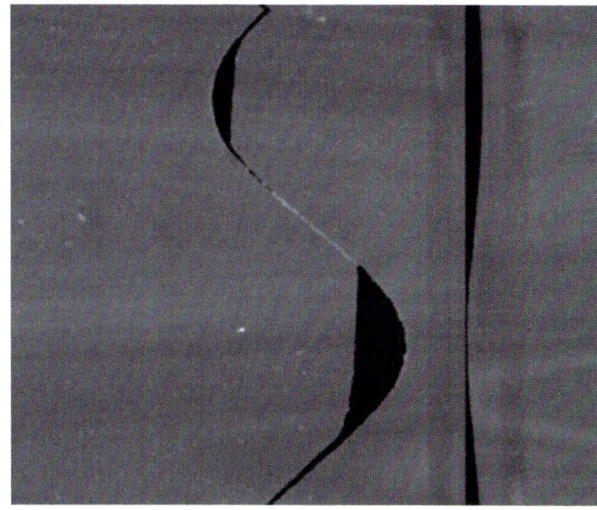

Figure 3F6.3 A CT scan of the core before it was removed from the barrel, showing a fracture cutting oblique to the core axis and the missing uphole and downhole tips of the core wedges. Horizontal 4-inch diameter core; uphole is to the left.

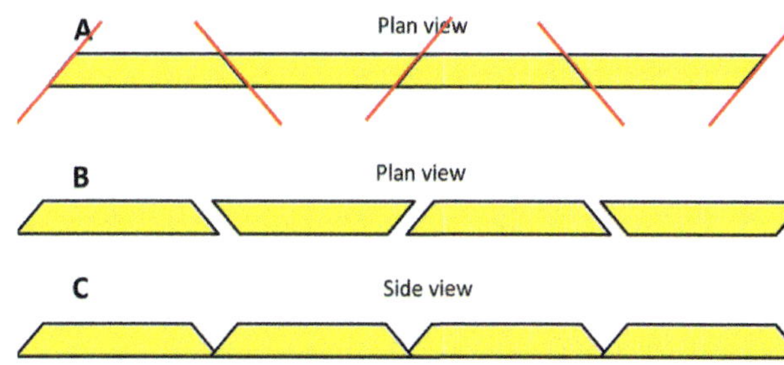

Figure 3F6.4 A: A horizontal core is cut through a formation containing two vertical fracture sets (red lines) that strike nearly normal to each other. B: While the lower parts of the core are being cut, this section of the core separates along the poorly mineralized fracture planes, with core segments shifting uphole into the unoccupied part of the core barrel. C: Due to gravity, the long, heavy sides of the trapezoidal core segments rotate 90° to lie at the bottom or low side of the core barrel, juxtaposing the tips of the core wedges, where they are ground away by coring-related vibration. When the core barrel is stood on end vertically as it is brought up the hole, gravity forces the core pieces back to their original positions across the disarticulated fracture planes.

Figure 3F6.5 The two pieces of core, separated by a calcite-mineralized fracture, rotated relative to each other from their in situ positions to positions that are gravitationally stable in a horizontal core barrel. The tips of the core were ground up and removed by vibration associated with the coring operation, which also imparted short, arc-shaped lineations to the tip-to-tip contact planes. Horizontal, 4-inch diameter core; stratigraphic up is towards the viewer in the core piece on the left and away from the viewer in the core piece on the right. Uphole is to the left.

3F7

Slab-Plane Consistency

Assuming both the slabs and butts of a slabbed core are available for study, most fracture analysts prefer to have both, and prefer to have the slab planes of a core cut so that they have the same orientation as much as possible along the core. However, if only the core slabs are available, inconsistent slab planes have a better chance of intersecting at least part of any long fracture than do consistently oriented slab planes, so there is some advantage to randomly oriented slabs. Many slabbing technicians are in fact instructed to cut slabs so that they will provide good images for photography, which often means that the natural fractures that can break up the core are intentionally avoided when choosing slab plane orientations (Figure 3F7.1), and are relegated to the butts of a core. Where only the core slabs or butts are available for study, inconsistently oriented slab planes commonly truncate and obscure parameters such as fracture height (Figure 3F7.2), strike (Figure 3F7.3), and number of fractures (Figure 3F7.4).

Figure 3F7.1 The slab face (white arrow) along this core was intentionally cut to avoid the 6-foot tall, calcite-mineralized natural fracture (red arrow) found only in the butt section of this shale core. Butts of a vertical, 4-inch diameter core; uphole is away from the viewer.

Atlas of Natural and Induced Fractures in Core, First Edition. John C. Lorenz and Scott P. Cooper.

Figure 3F7.2 Left: what appear to be two fractures in the slabs of a shale core in fact form one continuous mineralized fracture plane when the slabs and butts are reassembled. Right: the slab planes in this shale core are parallel, but the core piece immediately above the finger is upside down (shown as originally found in both the core box and in the photo logs from the service company), making it appear that there are three separate fractures. A fracture analyst must put hands on the core, picking up and examining all pieces, and fitting the pieces back together. Slabs from vertical, 4-inch diameter cores; uphole is towards the top of both photos.

Figure 3F7.3 Two views of an inclined extension fracture pair in core cut from a limestone. Left: a confusing fracture pattern is presented by the core as it is laid out in the slab box, due to a change in the slab-plane orientation across the break in the core. Right: the fracture pattern is in fact systematic and regular, but in order to make that obvious, the upper slab piece had to be rotated 180° around its long axis and mated to the opposing butt section of the core. Vertical, 4-inch diameter core; uphole is towards the top of both photos.

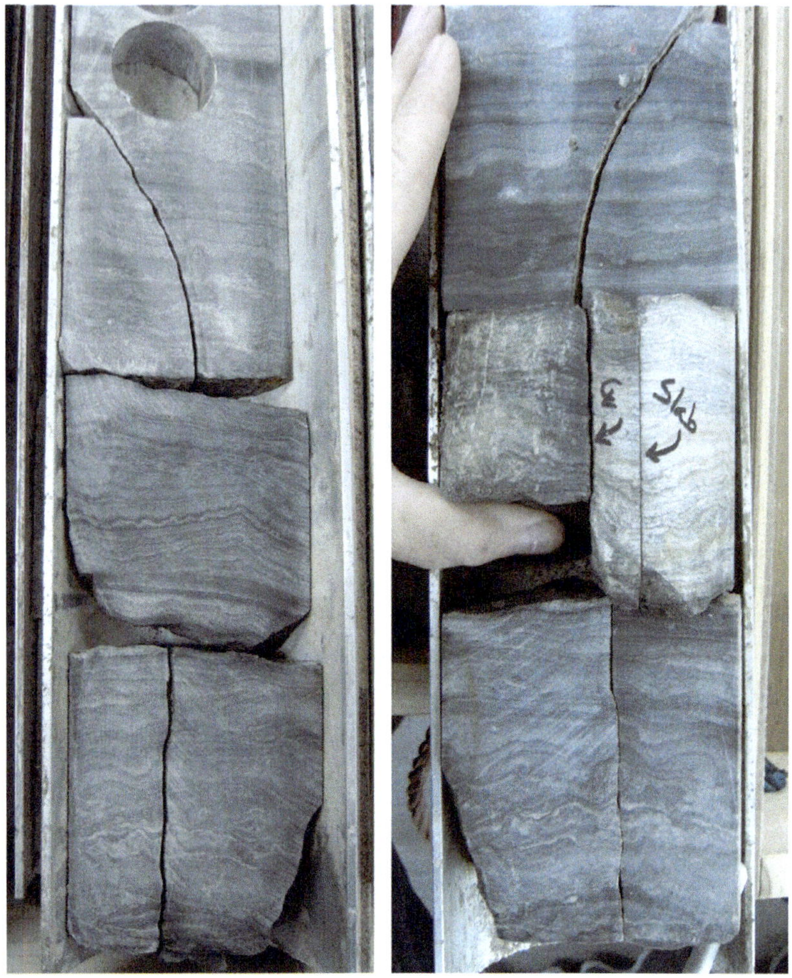

Figure 3F7.4 Two views of a petal-centerline fracture in core cut from a limestone. Left: as laid out in the slab box, there appear to be two unrelated fractures. Right: when the butt of the unfractured slab piece is incorporated into the puzzle, however, the centerline fracture ("CL") is shown to be continuous. The slab face in the seemingly unfractured core piece was cut parallel to the induced fracture. Vertical, 4-inch diameter core; uphole is towards the top of the photo.

3F8

Illusions

The study of fractures in core is a three-dimensional exercise. As in outcrops, the expression of a fracture on one 2-D surface provides only its apparent geometry. Wherever possible, fractures in cores can and must be traced from a slab surface onto the core ends, through to the backs of slabs, into the butts, and even into the walls of plug holes in order to characterize them correctly. Sometimes a core must be intentionally broken open in order to provide an important view of the third dimension of a fracture.

Core is a miniscule sample of the reservoir to begin with and a fracture study of only the slabs of a core misses three-quarters of the core volume. This significantly reduces the chance of characterizing cored fractures, in part because, unlike sedimentary structures, fracture planes do not fill the core volume, and there is no guarantee that a cored fracture will be exposed on the slab plane. Moreover, a study that considers only the faces of the slabs, or that uses only photos of the slab faces, is severely handicapped since even when they intersect, slab planes are rarely cut normal to fracture planes. Fracture planes oriented at oblique angles to the slab show distorted dimensions of the fracture (Figures 3F8.1–3F8.3); vertical fractures intersected by inclined planes appear to be inclined unless they strike exactly normal to the exposure plane (Figure 3F8.4).

Nevertheless, geology is a science of incomplete data sets, and although we invariably want more data, geologists have been quite successful at making the most of limited and incomplete data sets, and core photos are better than no data. One just needs to be aware of the limitations of such data, and this section illustrates some of the pitfalls of studying fractures only as they are expressed on a slab face.

Figure 3F8.1 Two views of a narrow, calcite-mineralized fracture in a muddy limestone. Left: the calcite-mineralized fracture above the yellow arrow appears to have a nearly horizontal dip angle, nearly parallel to bedding, as it is exposed on the slab plane. Right: examination of the back of the slab shows that in fact, the fracture has a high-angle dip. A near-vertical fracture striking parallel to the slab face can have a perfectly horizontal expression on the slab plane, a configuration that is no less probable than a vertical fracture that strikes with the ideal angle, perfectly normal to the slab. Vertical, 4-inch diameter core; uphole is towards the top of both photos.

Figure 3F8.2 Three views of fractures in a chalky marine limestone. Fractures that are cut nearly parallel to their strike by a slab plane appear to be not only wider but also much more irregular than they really are. Left: the irregular manifestation of the fracture on the slab surface. Right: two photos showing the ends of the core across the break in the middle of the left photo, illustrating the shallow angle between the fracture and the slab plane. Slab from a vertical, 3-inch diameter core; the core markings use an unconventional "down" arrow, and uphole is towards the top of both the left and lower right photos, and away from the viewer in the upper right photo.

Figure 3F8.3 Fracture irregularities are exaggerated when a slab face is cut across the fracture plane at a shallow angle, as in this core cut from a non-marine sandstone. The narrow fracture is nearly vertical (arrow, right photo), but has a nearly horizontal expression at the bottom of the left photo. Butt from a vertical, 4-inch diameter core; uphole is towards the top of the photo.

Figure 3F8.4 Left: three vertical, calcite-mineralized natural fractures ("NF") in a fine-grained limestone appear to be inclined because they are cut oblique to strike by an inclined petal-fracture plane ("PF"). The natural fractures even appear to form a conjugate fracture set with a vertical bisector to the acute intersection angle. However, 3D examination that includes their expression on the base of the core piece (right; the fractures are parallel to the black lines drawn on the core) shows that they are nearly vertical and have a right-angle intersection in map view. Vertical, 4-inch core; uphole is towards the top of the left photo, and away from the viewer in the right photo.

3F9

Coring-Related Rock Alteration on Core Surfaces

Cores are slabbed so that a better view of the sedimentary and structural features can be obtained than that offered by the rough and distressed outer core surfaces. The surfaces of some cores, particularly those cut from anhydrites and limestones, can be obscured by a rind of material consisting of mechanically and chemically altered rock created by the locally elevated heat, pressure, and stress during passage of the core bit (Figures 3F9.1–3F9.3). Similar alterations must occur in the wellbore, but their effects are usually assumed to be related to filter cakes of drilling mud that create low-permeability skins and that must be accounted for in logging and engineering operations.

Even when a core is clean, most sedimentary and stratigraphic features can be observed in the smaller, easier to examine and store slab pieces; core butts are underappreciated and some are even discarded. However, core butts contain about three-quarters of the volume of a core and provide more fracture information than slabs.

This brings up the common question: Should a core be logged for fractures before or after it has been slabbed? The answer, of course, is that it depends on the core condition and the nature of the cored fracture system. Slabbing can reveal fractures, especially the smaller ones, that are not visible if the core surface is obscured. However, no matter how carefully it is done, slabbing breaks up a core and degrades core continuity. Slabbed cores typically provide more information on parameters such as fracture distributions and widths but they provide less information on fracture orientations and heights. Clean cores from sandstones can often be logged for fractures before slabbing without missing much, but rough-surfaced muddy shale cores with small fractures may appear to be entirely unfractured until they are slabbed. Ideally, a core would be logged both before and after slabbing; we have had that luxury on a few projects and the two fracture data sets obtained from a core before and after slabbing are complementary but can be significantly different. Nevertheless, the most cost-effective fracture data sets are usually obtained from slabbed cores where the butts and slabs are logged together.

Atlas of Natural and Induced Fractures in Core, First Edition. John C. Lorenz and Scott P. Cooper.

Figure 3F9.1 Left: a textured, discolored, coring-related rind formed on a gray limestone. Right: a section across a similar core, along a calcite-mineralized natural fracture, showing the minimal thickness of the alteration rind. Vertical, 4-inch diameter core; uphole is towards the top of both photos.

Figure 3F9.2 The surface rind of this core, cut from a gray anhydrite, has been chemically altered and consists of white gypsum. Brown oil stains the gypsum in patches centered on the darker knots where the oil is bleeding from the core. Vertical, 4-inch diameter core; uphole is towards the top of the photo.

Figure 3F9.3 A rind of altered rock covers the surface of this limestone core, spalling off along the scribed orientation groove (right of center) but completely obscuring sedimentary structures and fractures alike elsewhere. The brown lines on the core at the left of the photo are stains left by the iron core-layout rack. Vertical, 4-inch diameter core; uphole is towards the top of the photo.

Index

Wiley The manufacturer's authorized representative according to the EU
General Product Safety Regulation is Wiley-VCH GmbH, Boschstr. 12,
69469 Weinheim, Germany, e-mail: Product_Safety@wiley.com.

Printed and bound by CPI Group (UK) Ltd, Croydon, CR0 4YY
15/09/2026
02217280-0003